MITIGATING SHORE EROSION ALONG SHELTERED COASTS

Committee on Mitigating Shore Erosion Along Sheltered Coasts

Ocean Studies Board

Division on Earth and Life Studies

NATIONAL RESEARCH COUNCIL
OF THE NATIONAL ACADEMIES

THE NATIONAL ACADEMIES PRESS
Washington, D.C.
www.nap.edu

THE NATIONAL ACADEMIES PRESS   500 Fifth Street, N.W.   Washington, DC 20001

NOTICE: The project that is the subject of this report was approved by the Governing Board of the National Research Council, whose members are drawn from the councils of the National Academy of Sciences, the National Academy of Engineering, and the Institute of Medicine. The members of the committee responsible for the report were chosen for their special competences and with regard for appropriate balance.

This study was supported by a contract between the National Academy of Sciences and the following entities: purchase order no. 2W-0373-NANX from the Environmental Protection Agency, purchase order numbers W19HQ-04-P-0132 and W912HQ-05-P-0064 from the U.S. Army Corps of Engineers, contract numbers 04-891 and 05-927 from the Cooperative Institute for Coastal and Estuarine Environmental Technology, and purchase order no. FC133CO5SE6428 from the NOAA Coastal Services Center. Any opinions, findings, conclusions, or recommendations expressed in this publication are those of the author(s) and do not necessarily reflect the views of the organizations or agencies that provided support for the project.

International Standard Book Number-13:   978-0-309-10346-6 (Book)
International Standard Book Number-10:   0-309-10346-0 (Book)
International Standard Book Number-13:   978-0-309-66651-0 (PDF)
International Standard Book Number-10:   0-309-66651-1 (PDF)
Library of Congress Catalog Card Number 20006941054

Additional copies of this report are available from the National Academies Press, 500 Fifth Street, N.W., Lockbox 285, Washington, DC 20055; (800) 624-6242 or (202) 334-3313 (in the Washington metropolitan area); Internet, http://www.nap.edu.

THE NATIONAL ACADEMIES
Advisers to the Nation on Science, Engineering, and Medicine

The **National Academy of Sciences** is a private, nonprofit, self-perpetuating society of distinguished scholars engaged in scientific and engineering research, dedicated to the furtherance of science and technology and to their use for the general welfare. Upon the authority of the charter granted to it by the Congress in 1863, the Academy has a mandate that requires it to advise the federal government on scientific and technical matters. Dr. Ralph J. Cicerone is president of the National Academy of Sciences.

The **National Academy of Engineering** was established in 1964, under the charter of the National Academy of Sciences, as a parallel organization of outstanding engineers. It is autonomous in its administration and in the selection of its members, sharing with the National Academy of Sciences the responsibility for advising the federal government. The National Academy of Engineering also sponsors engineering programs aimed at meeting national needs, encourages education and research, and recognizes the superior achievements of engineers. Dr. Wm. A. Wulf is president of the National Academy of Engineering.

The **Institute of Medicine** was established in 1970 by the National Academy of Sciences to secure the services of eminent members of appropriate professions in the examination of policy matters pertaining to the health of the public. The Institute acts under the responsibility given to the National Academy of Sciences by its congressional charter to be an adviser to the federal government and, upon its own initiative, to identify issues of medical care, research, and education. Dr. Harvey V. Fineberg is president of the Institute of Medicine.

The **National Research Council** was organized by the National Academy of Sciences in 1916 to associate the broad community of science and technology with the Academy's purposes of furthering knowledge and advising the federal government. Functioning in accordance with general policies determined by the Academy, the Council has become the principal operating agency of both the National Academy of Sciences and the National Academy of Engineering in providing services to the government, the public, and the scientific and engineering communities. The Council is administered jointly by both Academies and the Institute of Medicine. Dr. Ralph J. Cicerone and Dr. Wm. A. Wulf are chair and vice chair, respectively, of the National Research Council.

www.national-academies.org

COMMITTEE ON MITIGATING SHORE EROSION
ALONG SHELTERED COASTS

JEFF BENOIT, *Chair*, SRA International, Arlington, Virginia
C. SCOTT HARDAWAY, JR., College of William and Mary, Virginia Institute of Marine Science, Gloucester Point
DEBRA HERNANDEZ, Hernandez and Company, Isle of Palms, South Carolina
ROBERT HOLMAN, Oregon State University, College of Oceanic Atmospheric Sciences, Corvallis
EVAMARIA KOCH, University of Maryland, Center for Environmental Science, Horn Point Laboratory, Cambridge
NEIL McLELLAN, Shiner Moseley and Associates, Houston, Texas
SUSAN PETERSON, Teal Partners, Rochester, Massachusetts
DENISE REED, University of New Orleans, Department of Geology and Geophysics, New Orleans, Louisiana
DANIEL SUMAN, University of Miami, Rosenstiel School of Marine and Atmospheric Science, Miami, Florida

Staff

SUSAN ROBERTS, Study Director
AMANDA BABSON, Christine Mirzayan Science and Technology Policy Graduate Fellow
SARAH CAPOTE, Senior Program Assistant

Preface

Sheltered coasts, and the bodies of water they surround, are increasingly popular places for people to live, work, and recreate. This is partially due to a preference for a serene setting that is afforded some protection from the full fury of coastal storms. The same sheltered nature of these areas also creates some of the most biologically productive and ecologically valuable resources of the coastal region. But sheltered coasts are indeed subject to erosion and sea level rise, and suffer chronic land loss as a result. The common response by landowners to the loss of their increasingly valuable land has been to "harden" the shoreline by construction of fixed structures such as bulkheads, revetments, or groins. This response can easily create a "bathtub" effect and fundamentally change the character of the coastal environment; in some cases actually worsening the erosion and inundation, and in most cases causing a loss or shift in ecological values. As a better understanding is gained of the physical and ecological impacts of hardening the shoreline, new approaches are being developed for managing eroding sheltered shorelines. This report reviews options available to mitigate erosion of sheltered coasts; explores why certain decisions are made regarding the choice of erosion mitigation options; provides critical information about the consequences or altering sheltered shorelines; and, provides recommendations about how to better inform decisions in the future.

Integrating broad societal and ecological considerations into erosion mitigation strategies is a continuing challenge. Practitioners are slowly moving in that direction and are encouraged to continue on that course. We suggest that more focused research on sheltered coasts, and long-term regional planning early in the process, are key solutions to this chronic issue.

These findings would not have been possible without the hard work, collective action, and perseverance of this committee. I would like to thank my colleagues on the committee for their efforts. Some members are knowledgeable practitioners with hands-on experience working with landowners on the design or permitting of erosion mitigation strategies; others are among the leading researchers in the field of estuarine ecology. I have been honored to serve with such an eminent group and to learn from their wisdom. More importantly, it was a pleasure to get to know them and work alongside them all.

The committee and I gratefully acknowledge and thank the staff of the Ocean Studies Board for their tireless support. Dr. Susan Roberts served as project director for the majority of the project after Dr. Dan Walker was called on for service to another committee. Dan was invaluable for getting us off to a well thought out start and assisting with preparation of the final report. Frank Hall and Susan Park provided valuable assistance during the review process of this report. Sarah Capote was always there for us and provided critical research and logistical support. Amanda Babson also conducted helpful research for the workshop during her internship with the Board.

We hope that the conclusions and recommendations of this report provide meaningful advice to state and federal agencies, local communities, and landowners to guide a new management approach in the planning, design, and construction of erosion mitigation strategies for sheltered coasts.

Jeffrey R. Benoit, *Chair*

Acknowledgments

This report was greatly enhanced by participants at the two meetings held as part of this study. The committee would like first to acknowledge the efforts of those who gave presentations at the meetings. These talks helped set the stage for fruitful discussions in the closed sessions that followed.

ROBERT BRUMBAUGH, U.S. Army Corps of Engineers
BETH BRYANT, University of Washington, School of Marine Affairs
CHARLES CHESTNUTT, U.S. Army Corps of Engineers
SCOTT DOUGLASS, University of South Alabama
KATHLEEN KUNZ, U.S. Army Corps of Engineers
ROBIN LEWIS, Lewis Environmental Services, Inc.
DAN MILLER, New York State Department of Environmental Conservation, Hudson River National Estuarine Research Reserve
DOUG MYERS, Puget Sound Action Team
KAREN NOOK, U.S. Army Corps of Engineers
NEVILLE REYNOLDS, Vanasse Hangen Brustlin, Inc.
SPENCER ROGERS, North Carolina Sea Grant
HUGH SHIPMAN, Washington Department of Ecology
JAY TANSKI, New York Sea Grant
JOHN TEAL, Teal Partners and Woods Hole Oceanographic Institution
JIM TITUS, Environmental Protection Agency
DWIGHT TRUEBLOOD, National Oceanic and Atmospheric Administration
DON WARD, U.S. Army Corps of Engineers
PHIL WILLIAMS, Phillip Williams Associates, Ltd.

The committee is also grateful to **Hugh Shipman**, Washington Department of Ecology, and **Jeff Parsons**, University of Washington, for organizing the committee field trip in the Seattle area. Assistance with conceptual diagrams in Chapter 4 was provided by the Integration and Application Network (ian.umces. edu), University of Maryland Center for Environmental Science.

This report has been reviewed in draft form by individuals chosen for their diverse perspectives and technical expertise, in accordance with procedures approved by the National Research Council's Report Review Committee. The purpose of this independent review is to provide candid and critical comments that will assist the institution in making its published report as sound as possible and to ensure that the report meets institutional standards for objectivity, evidence, and responsiveness to the study charge. The review comments and draft manuscript remain confidential to protect the integrity of the deliberative process. We wish to thank the following individuals for the participation in their review of this report:

BERNARD O. BAUER, University of British Columbia-Okanagan, Kelowna, Canada
MARK BRINSON, East Carolina University, Greenville, North Carolina
ROBERT DALRYMPLE, Johns Hopkins University, Baltimore, Maryland
MARGOT GARCIA, Tucson, Arizona
JORDAN LORAN, Maryland Department of Natural Resources, Annapolis
HUGH SHIPMAN, Washington Department of Ecology, Bellevue
CLIFF TRUITT, Coastal Technology Corporation, Sarasota, Florida
S. JEFFRESS WILLIAMS, U.S. Geological Survey, Woods Hole, Massachusetts

Although the reviewers listed above have provided many constructive comments and suggestions, they were not asked to endorse the conclusions or recommendations nor did they see the final draft of the report before its release. The review of this report was overseen by **Norbert Psuty**, Rutgers University. Appointed by the National Research Council, the reviewers were responsible for making certain that an independent examination of this report was carried out in accordance with institutional procedures and that all review comments were carefully considered. Responsibility for the final content of this report rests entirely with the authoring committee and the institution.

Contents

Summary

Erosion is a natural phenomenon that threatens the growing number of properties built on coastal shores. Although open coasts have been the focus of most studies on erosion and technologies for stabilizing the shoreline, sheltered coastal areas,[1] such as those found in bays and estuaries, also suffer land loss from erosion and high waters. For example, the Maryland Geological Survey estimated that Maryland lost more than 20 acres of land on the western shore of Chesapeake Bay in the wake of Tropical Storm Isabel, causing $84,000,000 in damages to shoreline structures (Maryland Department of Planning, 2004, Baltimore, MD. 29 pp. *Lessons learned from Tropical Storm Isabel*).

Landowners frequently respond to the threat of erosion by armoring the shoreline with bulkheads, revetments, and other structures. Although the armoring of a few properties has little impact, the proliferation of structures along a shoreline can inadvertently change the coastal environment and the ecosystem services that these areas provide. Managers and decision-makers have been challenged to balance the trade-offs between protection of property and potential loss of landscapes, public access, recreational opportunities, natural habitats, and reduced populations of fish and other living marine resources that depend on these habitats.

At the request of the Environmental Protection Agency, the Army Corps of Engineers, and the Cooperative Institute for Coastal and Estuarine Environmental Technology, this report examines the impacts of shoreline management on sheltered coastal environments and strategies to minimize potential negative

---

[1] A glossary of terms used in this report can be found in Appendix E.

impacts to adjacent or nearby coastal resources (see Box S-1). The report suggests the development of a new shoreline management framework that would help decision makers evaluate the spectrum of available approaches to shoreline erosion problems in the context of the environmental setting, including the physical properties and ecological services of sheltered coasts and the potential for cumulative impacts.

SHELTERED COASTS AND EROSION

Sheltered coasts are shorelines that face smaller bodies of water in comparison to the beaches found facing the open ocean. Similarly, lagoons formed by fringing coral reefs or sand bars, which experience reduced wave energy, have relatively protected shorelines. Many of the processes that govern erosion on the open coast also apply to sheltered coasts, but compared to the typically long linear nature of open coasts, sheltered shorelines exhibit a more irregular configuration and often display very distinct geomorphic compartments containing a complex mix of resources that may vary from compartment to compartment. The lower energy conditions found on sheltered coasts create unique environments that foster habitats and ecological communities, such as marshes and mudflats, typically not found on open coasts. The differences between the shore dynamics and habitats on sheltered versus open coasts affect the potential technological approaches and the consequences of actions taken to stem erosion and land loss from sea-level rise.

Erosion is a natural phenomenon caused by winds, waves, currents, and tides that pick up and transport shoreline sediments; and weathering processes that destabilize landforms such as dunes and bluffs. Although natural processes contribute to erosion, the rate may be accelerated by human activities such as construction of dams upstream of estuaries or rivers that trap sediments, or installation of groins and seawalls that alter the magnitude and direction of sediment transport. Similarly, inundation may increase if land subsides due to natural compaction of sediments or due to withdrawal of subsurface resources, such as groundwater and petroleum. Other human activities that increase erosion include dredge and fill operations, wetland drainage, boat traffic, and channel dredging.

Superimposed on the impacts of erosion and subsidence, the effects of rising sea level will exacerbate the loss of waterfront property and increase vulnerability to inundation hazards. Sea level rise changes the location of the coastline, moving it landward along low lying contours and exposing new areas and landforms to erosion. Additionally, sea-level rise is chronic and progressive, requiring a response that is correspondingly progressive. Attempts to follow a "hold the line" mitigation[2] strategy against erosion and sea-level rise by coastal armoring

---

[2]In this report, "erosion mitigation" is used to describe efforts to reduce or lessen the severity of erosion and should not be confused with mitigation of environmental damages.

> ## BOX S-1
> ## Statement of Task
>
> The study will examine the impacts of shoreline management on sheltered coastal environments (e.g., estuaries, bays, lagoons, mudflats, deltaic coasts) and identify conventional and alternative strategies to minimize potential negative impacts to adjacent or nearby coastal resources. These impacts include: loss of intertidal and shallow water ecosystems, effects on species, and loss of public trust uses. The study will provide a framework for collaboration between different levels of government, conservancies, and property owners to aid in making decisions regarding the most appropriate alternatives for shoreline protection.
> In particular, the committee will address the following questions:
>
> • What engineering techniques, technologies, and land management/planning measures are available to protect sheltered coastlines from erosion or inundation resulting from either natural or anthropogenically forced processes? When does the design and implementation of these measures require a distinction between natural and anthropogenic causes and how can this be achieved?
> • What information is needed to determine where and when these measures are reliable and effective both from an engineering and a habitat perspective? What are the likely individual and cumulative impacts of shoreline protection practices or no action on sheltered coastal habitats including public and private property, and public access along the shore, locally and regionally?
> • Over what time frame are monitoring data needed to document the effectiveness of protective coastal measures? What data are needed to predict when design criteria may be exceeded?
> • Given current trends in erosion and inundation rates and a possible acceleration of relative sea-level rise, how can design criteria, the mix of technologies employed, and land use plans be implemented for the protection of the environment and property over the long term?

will result in a steady escalation in both the costs of maintenance and the consequences of failure.

CURRENT APPROACHES TO MITIGATE EROSION

The pressure to develop and stabilize shorelines in sheltered coastal areas is increasing; more people desire waterfront homes, raising coastal property values and creating strong incentives to protect high-priced real estate. There are several types of mitigation measures to stabilize shorelines, including structural hardening (e.g., seawalls, bulkheads, revetments) and alternatives, such as constructed marsh fringes, that are designed to preserve a more natural shoreline. The selection of the type of response to prevent or offset land loss depends on understanding local causes of erosion or inundation.

The most common response to erosion of sheltered shorelines has been a "hold the line" strategy that relies on technologies that harden the shoreline. A shift away from this approach has been slow, in part because there is a greater familiarity with these methods than with alternative approaches such as constructing a marsh fringe or using vegetation to stabilize a bluff. Contractors are more likely to recommend structures such as bulkheads because they have experience with the technology and know the design specifications and expected performance. Landowners expect that a hard, barrier-type structure will be required to prevent loss of property and protect buildings. In many regions, the regulatory system may unintentionally encourage shoreline armoring because it is simpler and faster to obtain the required permit(s).

However, there are indirect costs associated with mitigation options that armor the shoreline, including loss of ecosystem services at the site and in surrounding waters and shorelines. Many of these costs are borne by the public rather than the landowner. For example, installation of a seawall can result in loss of the fronting beach with attendant loss of public access and scenic amenities. Seawalls and bulkheads may also lead to loss of the intertidal zone and an exchange of habitat types from soft to hard substrates with subsequent changes in the plants and animals that inhabit these areas. When marshes are lost as the result of an installation, a highly diverse and productive plant and animal community is lost with attendant loss of vital ecosystem services such as nursery areas for important fish stocks, removal of excess nutrients from land runoff, feeding areas for migratory birds, and sediment stabilization. Some types of armoring may affect erosion patterns in nearby areas through scouring at the edges of structures or through disrupting the transport of sediment to downstream areas.

A NEW SHORELINE MANAGEMENT FRAMEWORK

Changing the current practice of armoring sheltered coasts will require a change in the shoreline management framework. Decision-makers should appreciate the costs and benefits of the spectrum of potential solutions to shoreline erosion problems, including potential cumulative impacts on shoreline features, habitats, and other amenities. The management framework should encourage approaches that minimize habitat loss and enhance natural habitats in environments where such methods offer effective stabilization.

Overcoming the obstacles associated with the existing management framework will require a number of societal and institutional changes that include:

• better understanding of sheltered shoreline processes and ecological services,

- improved awareness of the choices available for erosion mitigation,
- documentation of individual and cumulative consequences of erosion mitigation approaches,
- shoreline management planning that takes into consideration the unique ecological and physical processes of sheltered coasts, and
- a permitting system with incentives that support the goals of the shoreline management plan.

The study's main findings and recommendations on these points are summarized below.

Understanding Sheltered Shoreline Processes and Ecological Services

Overall, less is known about physical process of sheltered coastal systems than of open coasts. Basic information, such as resource characterization, shoreline change analysis, sediment transport patterns, habitat function, and ecological services, is available for only a portion of the nation's sheltered shorelines and few programs address these knowledge gaps. States generally lack sufficient resources to conduct the type of comprehensive assessment of shorelines required for effective regional planning. However, decision-makers need adequate information about the physical and biological systems that will be affected to make well-informed choices about erosion mitigation along sheltered coasts.

FINDING: In most areas, the scope and accessibility of information regarding the causes of erosion at specific sites and the overall patterns of erosion, accretion, and inundation in the broader region (estuary, lagoon, littoral cell) is insufficient to support the development of an integrated plan for managing shore erosion.

RECOMMENDATIONS:
- **Federal agencies (e.g., USACE, EPA, USGS, and NOAA), state agencies, and coastal counties and communities should support targeted studies of sheltered coast dynamics to provide an informed basis for selecting erosion mitigation options that consider the characteristics of the broader coastal system rather than simply addressing immediate problems at individual sites. These studies should:**

— **Identify trade-offs in ecosystem services associated with various mitigation measures,**
— **Quantify the costs and benefits of nonstructural erosion control techniques,**
— **Document system-wide process and hazard information, including mapping of erosion zones and rates. This information needs to be**

presented in non-technical formats such as summary maps that can be readily understood by decision-makers, and
— Develop models to predict the evolution of coastal features under various scenarios.

• State and federal agencies should ensure that the information obtained from these studies is readily available to the public and decision makers at all levels of government.

Improved Awareness of the Choices Available for Addressing Erosion

One barrier to changing the trend towards increased shoreline armoring is a general lack of knowledge and experience among decision-makers, particularly property owners, regarding alternative options for shoreline erosion response, the relative level of erosion mitigation afforded by the alternative approaches and their expected lifetime, and the nature of the associated impacts and benefits. This unfamiliarity with alternative engineering approaches has resulted in disinterest, concern, or disagreement among regulators regarding the ecological consequences of alternative shoreline stabilization measures.

FINDING: Many decision-makers, particularly homeowners but also some state and federal regulators, are not sufficiently informed about the mitigation options available to them or the short and long term impacts of their choices. Decision-makers need assessments of new techniques and materials designed to mitigate shore erosion. Because of the comparatively low energy environments on sheltered coasts, special techniques have been developed to address erosion in these areas. Some techniques, such as the combination of a planted marsh fringe with a sill, have been tested and proven effective under well-characterized physical settings. However, new techniques (or structural materials) are periodically introduced that require a rigorous process of testing and evaluation to determine their effectiveness in controlling erosion and to evaluate their impacts on the environment.

RECOMMENDATIONS:
• **The major federal agencies involved in permitting activities (EPA, USACE, and NOAA) should initiate a national policy dialogue on erosion mitigation for sheltered coasts to bring together state and federal decision-makers and share information on the potential application and value of different mitigation approaches.**
• **The national dialogue should be used to develop guidelines for mitigating erosion on sheltered coasts that give deference to ecologically beneficial measures and ensure consistency of decision-making across regions.**

- As part of the national dialogue, the permitting agencies should develop publications that contain objective information about erosion mitigation techniques, including descriptions of the conditions under which each option would be effective. These publications (either print or electronic) should be actively distributed to state and local planning and permitting staff, professional associations of environmental consultants, engineers, zoning officials, planners, and building inspectors; and extension agents; and made readily available to property owners and community groups.
- Professional societies and conferences should be utilized as a venue for transferring information to decision-makers such as regulators, engineers, and consultants.

Cumulative Consequences of Erosion Mitigation Approaches

Cumulative impacts[3] encompass the combined effects on legal, social, ecological, and physical systems. From a legal or regulatory perspective, issuance of a permit may establish a precedent, potentially facilitating the approval process for future requests for similarly situated structures. Another aspect of cumulative impact is the erosion enhancing effect of structures such as bulkheads on adjoining properties. Flanking property owners are likely to respond by constructing their own bulkheads, with a domino-type effect up and down the shoreline. It is difficult to identify the point at which individual projects accumulate to an extent that threatens the valued properties of the shoreline.

FINDINGS:
- Although loss of small parcels of shoreline habitat from hardening may not have a large impact on the ecosystem, the cumulative impact of the loss of many small parcels will at some point alter the properties, composition, and values of the ecosystem. In addition, the economic, recreational, and aesthetic properties of the shoreline will be altered, with potential loss of public use, access, and scenic values.
- Cumulative effects of shoreline hardening projects are rarely assessed and hence are generally unknown. However, an appreciation of the potential cumulative effects will be necessary to prevent an underestimation of the impacts of individual projects.

---

[3]Cumulative impacts are defined as "the impacts on the environment, which result from the incremental impact of the action when added to other past, present and reasonably foreseeable future actions regardless of what agency (Federal or non-Federal) or person undertakes such other actions" (40 CFR 1508.7 and 1508.8).

RECOMMENDATIONS:
• Cumulative effects should be considered in shoreline management plans, both for the values invested by the affected communities in non-hardened shorelines and the value of ecosystem properties that stand to be lost with shoreline hardening. Although it may not be possible to identify the threshold beyond which cumulative impacts become unacceptable or irreversible, anticipation of the problem allows prioritization of projects in areas unsuited to nonstructural alternatives or sites where structures are predicted to have less impact.
• In the absence of a comprehensive assessment of the cumulative impacts of erosion mitigation measures, a precautionary approach should be used to prevent the unintentional loss of shoreline features and significant alteration of the coastal ecosystem.

Permitting System

FINDINGS:
• The current permitting system fosters a reactive response to the problem of erosion on sheltered coasts. Decision-making is usually parcel-by-parcel and based on little or no physical or ecological information. The path of least resistance drives choices through a rigid decision-making process, with inadequate attention to the cumulative effects of individual decisions.
• The current regulatory framework for sheltered coasts contains disincentives to the development and implementation of erosion control measures that preserve more of the natural features of shorelines, mainly as a result of the combined lack of knowledge, vision, and planning.

RECOMMENDATIONS:
• State and federal agencies (EPA, USACE, and NOAA) need to convene a working group to evaluate the decision-making process used for issuing permits for erosion mitigation structures to revise the criteria for sheltered coasts, including consideration of potential cumulative impacts.
• The regulatory preference for permitting bulkheads and similar structures should be changed to favor more ecologically beneficial solutions that still help stabilize the shore.
• State and federal regulatory programs (or other programs as appropriate) should establish a technical assistance function to provide advice on permitting issues and information on types of erosion mitigation approaches and their effectiveness under various site conditions.

Shoreline Management Planning

Creating a more proactive "regional approach" to shoreline management could address the unintended consequences of reactive permit decisions. The term "regional" is used in this report to reflect an area of shoreline that is defined by functional physical or ecological parameters such as littoral cells, or the scale of processes that affect sediment transport. Several examples of regional planning already exist for shorelines: the USACE Regional Sediment Management (RSM) approach, the EPA National Estuary Program, and some special area management plans approved by state coastal management programs.

FINDINGS:
• **The RSM approach provides a model and framework that could be adapted to address sheltered shoreline erosion problems within a regional context. Many factors in addition to sediment budgets must be considered in the development of regional shoreline management plans. These factors include socioeconomic considerations (e.g., ownership of the shoreline, waterfront property values, beach access for recreational boating and fishing) as well as a broad range of habitat and other ecological issues.**
• **Regional plans facilitate the assessment of cumulative impacts but require credible monitoring of project performance within and without the region of interest. Regional shoreline management plans could be created under the auspices of the Federal Coastal Zone Management Act (CZMA), Section 309 - Special Area Management Plans, thereby providing an opportunity to employ the federal consistency provisions of the CZMA to ensure that federal permitting actions are consistent with the plan.**

RECOMMENDATIONS:
• **Proactive erosion mitigation plans should be implemented to avoid unintended consequences when hardened shorelines reduce the recreational, aesthetic, economic, and ecological value of sheltered coastal areas.**
• **The essential elements of a regional shoreline management plan should include: (1) a shared vision for the future shoreline of the water body through stakeholder collaboration, (2) analysis of regional sediment budgets and the cumulative effects of existing shoreline management activities, (3) the mechanism for turning the vision into reality through consistent permitting provisions, (4) implementation, and (5) performance evaluation and monitoring requirements.**
• **Plans should be considered "living documents" and updated every 5 to 10 years as new information (e.g., monitoring data, research results) becomes available.**
• **Each regional shoreline management plan should describe the physical and hydrodynamic settings, including the location and type of existing**

shoreline structures in a Global Information System (GIS) format. The plan should describe the available mitigation options and discuss the applicability, relative cost and benefit, and effectiveness of each option.

• Monitoring should include both a preconstruction baseline and more detailed assessments after project implementation, both at the individual project level and for the entire region covered by the plan. Individual monitoring should be the responsibility of project proponents while regional monitoring should be the responsibility of the management plan authority.

• Regional shoreline management plans (based on estuary, bay, or littoral cell as appropriate) should be developed by local, state, and federal partners to address erosion on sheltered shorelines in a comprehensive, proactive manner.

• Information obtained from monitoring programs should be incorporated in subsequent planning activities to support adaptive management as a mechanism to consistently evaluate and refine regional plans.

CONCLUSION

Until the regulatory framework addresses the regional scale of the processes controlling sediment transport, stabilization of individual sites will often include structures that damage adjacent areas and create a domino-type effect of coastal armoring. Therefore, the dimensions of the regulatory framework should match the scale of the processes that contribute to shore erosion.

Currently there is no national mandate to document erosion processes on sheltered coasts or to develop regional scale plans. No federal agency has been assigned to provide that scale of planning, although some states have become more proactive in shoreline management. Hence, implementation of a regional plan will require a new commitment for coordination among local, state, and federal programs, including a regional general permit.

1

Introduction

Coastlines are perpetually changing—some from natural processes—some from human activities—many from both. The frequent human response to erosion is an attempt to stabilize, or "harden," the shoreline. Usually this is an approach that results in long-lasting consequences for the natural system, not just locally but also affecting surrounding areas. There are however many effective alternatives to hardening and depending on the selections made, the long-term consequences to the area can be positive or negative. This report reviews options available to address and mitigate[1] erosion of sheltered coasts and explores why certain decisions are made regarding the choice of erosion mitigation options; provides critical information about the consequences of altering sheltered shorelines; and, provides recommendations about how to better inform decisions in the future.

STUDY HISTORY

Before decisions can be made concerning appropriate shoreline management strategies on sheltered coasts, several topics must be understood. These include:

- which natural and anthropogenic factors are responsible for land losses, and how they occur;

---

[1]In this report, "erosion mitigation" is used to describe efforts to reduce or lessen the severity of erosion and should not be confused with mitigation of environmental damages.

- why present protection techniques and planning strategies are failing, and what impacts they have had;
- alternative protection techniques and their best environmental design criteria;
- how monitoring and data collection can be used to increase the effectiveness of protection strategies on developed and underdeveloped shorelines;
- which planning solutions exist that protect the environment while still allowing economic development including an understanding of the potential outcome of a "no action" decision; and
- how to provide federal and state agencies, local officials, managers, and the private sector with a framework for collaboration to avoid ill designed solutions, and instead obtain integrated, long-term, effective shoreline management.

To help address some of these questions, the U.S. Environmental Protection Agency's (EPA)[2] Global Programs Division first approached the Ocean Studies Board about developing a study on erosion mitigation options for sheltered coastlines. With additional interest and funding from the Army Corps of Engineers (USACE), the National Oceanic and Atmospheric Administration (NOAA)-Cooperative Institute for Coastal and Estuarine Environmental Technology (CICEET), and NOAA Coastal Services, the Ocean Studies Board assembled a committee of experts[3] to undertake a study defined by the statement of task given in Box 1-1.

SCOPE OF THE PROBLEM

Throughout the coastal regions of the world there are a significant number of areas that are partially or fully protected from the high-energy regimes associated with open coastlines, such as ocean-facing beaches. The sponsors requested that this study examine sheltered coastal environments such as estuaries, bays, lagoons, mudflats, and deltaic coasts (Box 1-1). These environments may be generally characterized as lower energy coastlines, but there is no quantitative formula that covers the diversity of conditions encountered on these more protected shorelines. Sheltered shorelines, though often contiguous with the open coast, border smaller, contained, bodies of water separated from the open ocean by islands, peninsulas, reefs, or other geomorphic features. Many of the processes that govern erosion and deposition on the open coast also apply to sheltered coasts, but generally the scale at which these processes function is significantly reduced within sheltered coastal areas. Also, unlike the typically long linear features associated with open coasts, sheltered coasts exhibit characteristics that are distinctively more compartmentalized with discrete areas of the coast

---

[2]For a list of acronym definitions, see Appendix B.
[3]For committee and staff biographies, see Appendix A.

BOX 1-1
Statement of Task

The study will examine the impacts of shoreline management on sheltered coastal environments (e.g., estuaries, bays, lagoons, mudflats, deltaic coasts) and identify conventional and alternative strategies to minimize potential negative impacts to adjacent or nearby coastal resources. These impacts include: loss of intertidal and shallow water ecosystems, effects on species, and loss of public trust uses. The study will provide a framework for collaboration between different levels of government, conservancies, and property owners to aid in making decisions regarding the most appropriate alternatives for shoreline protection.

In particular, the committee will address the following questions:

• What engineering techniques, technologies, and land management/planning measures are available to protect sheltered coastlines from erosion or inundation resulting from either natural or anthropogenically forced processes? When does the design and implementation of these measures require a distinction between natural and anthropogenic causes and how can this be achieved?

• What information is needed to determine where and when these measures are reliable and effective both from an engineering and a habitat perspective? What are the likely individual and cumulative impacts of shoreline protection practices or no action on sheltered coastal habitats including public and private property, and public access along the shore, locally and regionally?

• Over what time frame are monitoring data needed to document the effectiveness of protective coastal measures? What data are needed to predict when design criteria may be exceeded?

• Given current trends in erosion and inundation rates and a possible acceleration of relative sea-level rise, how can design criteria, the mix of technologies employed, and land use plans be implemented for the protection of the environment and property over the long term?

encompassing a variety of geomorphic and biological resources. Chapter 2 of this report describes some of the typical physical conditions associated with sheltered coasts; relatively low velocity tidal currents and mid-to-low energy wave climates associated with a limited fetch (distance from shore to shore). These conditions promote the formation of ecological complexes (i.e., mangroves, marshes, and mudflats) that often characterize habitats on sheltered coasts and are generally not found along open coasts.[4]

Many of these sheltered areas are river valleys that have been drowned by rising sea-level, or drainage features that are protected by headlands or islands. Many of these semiprotected sheltered shorelines border estuaries. It is estimated

---

[4]See "Terminology" section of this chapter for a more detailed discussion on the definition of a sheltered coast.

that in the United States alone there are 850 estuaries representing over 80 percent of the Atlantic and Gulf of Mexico coastal areas and more than 98 percent of the Virginia and Maryland coastlines (Nordstrom, 1992). Although estuaries are recognized as some of the most biologically productive areas of coastal regions, they are also increasingly popular places for people to live, work, and recreate. Estuaries are complex and dynamic systems subject to natural processes associated with wind, wave, and tidal action, which in turn cause erosion and subsequent transport and deposition of sediments along the shorelines. In addition to natural erosion, some anthropogenic activities such as recreational boating and commercial shipping can also contribute to erosion and sediment movement. When erosion occurs in the same area where human induced development exists, a "problem" is perceived and actions are taken to prevent the erosion. Historically the common practice to deal with erosion has been to use a variety of structures designed in some way to interrupt the natural processes of erosion, sediment transport and/or deposition. Such efforts may or may not be successful at preventing erosion, but most interrupt natural processes, frequently causing erosion in other areas and loss of the original habitats and other natural shoreline features.

The following case studies illustrate two examples of the growing problem of eroding sheltered shorelines and the severity of how human responses have altered the function of sheltered shorelines.

Mobile Bay, Alabama

Mobile Bay is a shallow body of water (average depth about 3 meters [approx. 10 feet]) that empties into the Gulf of Mexico. The bay is approximately 51 kilometers (about 32 miles) long, 37 kilometers (about 23 miles) across at its widest point, and the overall length of the shoreline is approximately 160 kilometers (about 100 miles). The bay area experiences extratropical storms and is subject to hurricanes from the Gulf of Mexico. With an average flow of 1,800 cubic meters (approx. 62,000 cubic feet) per second, the Mobile Bay estuary has the fourth largest freshwater flow in the continental United States (MBNEP, 2002). In 1997, a study was conducted that investigated the effects of bulkheads on the bay shoreline (Douglas and Pickel, 1999). The study examined aerial photographs from 1955, 1974, and 1985 to document the extent of bulkheads around the bay. Aerial video and site visits were used in 1997 to document the "present conditions." Results of the study reveal that the amount of armoring of the shoreline increased dramatically from 8 percent in 1955 to 30 percent in 1997. Table 1-1 is a summary of the length of shoreline armoring.

The study also documented that the rate of armoring from 1955 to 1997 corresponded to the rate of population growth in the area. This is a strong, but not surprising indication, that development trends increase the human impacts on sheltered coastlines. Of greater concern than the amount or rate of armoring was the associated loss of intertidal habitat, roughly estimated at 4 to 8 hectares

TABLE 1-1 Length of Shoreline Armoring Along Mobile Bay, Alabama

Year	Armored Shoreline	Natural Shoreline
1955	12,200 meters	145,000 meters
	(approx. 39,900 feet or 8%)	(approx. 475,600 feet or 92%)
1974	22,000 meters	135,200 meters
	(approx. 72,000 feet or 14%)	(approx. 443,500 feet or 86%)
1985	40,200 meters	116,900 meters
	(approx. 132,000 feet or 26%)	(approx. 383,500 feet or 74%)
1997	46,760 meters	110,400 meters
	(approx. 153,400 feet or 30%)	(approx. 362,100 feet or 70%)

SOURCE: Douglass and Pickel (1999).

(approx. 10 to 20 acres) corresponding to about 10 kilometers (approx. 6 miles) of intertidal beach shoreline.

Raritan Bay, New Jersey

Raritan Bay extends from the mouth of New Jersey's Raritan River to the entrance to the Atlantic Ocean, between the Verrazano Narrows to the north and Sandy Hook to the south. It measures about 22 kilometers (approx. 12 nautical miles) in length, and is about 13 kilometers (approx. 8 miles) wide at its widest point. The bay has an average depth of between 2 and 3 meters (7 and 10 feet), a tidal range of 1 to 2 meters (5 to 6 feet), and currents tend to flow east-west at about .5 to 1.5 knots. The shoreline of the bay is extensively altered by a variety of human development and stabilization efforts. The bay area is subject to frequent extratropical storms and periodic hurricanes.

A 14-kilometer (approx. nine mile) segment of the New Jersey side of the Raritan Bay shoreline was examined by Jackson (1996) for the types of alteration that have historically occurred to the shoreline in the presence of development and how they affected the sandy beaches. The shoreline of the study area is characterized by differing combinations of eroding bluffs, narrow beaches, and marsh; recreational and commercial development; and shore stabilization including groins, bulkheads, beach nourishment. A variety of analytical approaches were used to evaluate changes over time of development trends, beach and marsh environments, and shoreline positions.

Prior to the 19th century the area consisted largely of low bluffs and salt marsh with limited commercial and residential development. Development began in earnest in the late 19th century as a summer resort community for nearby New York City. Marsh areas were filled as homes, bulkheads, boardwalks and piers began to be constructed. Today the area is almost completely developed, predominately with year-round residences. Attempts to alter the shoreline have consisted of the use of shore parallel bulkheads and seawalls, shore perpendicular groins, and beach fill (nourishment). The use of bulkheads and seawalls increased over time and are the most common structures found along the study area. The use of groins was common prior to the 1950s but their use has diminished and few have been constructed since the 1970s. Surprising for a bay shoreline, the use of beach fill has been used extensively, although largely locally, in the study area. It was estimated that 2.8 million m^3 (approx. 3.7 million yd^3) of sand has been placed on the beaches, generally for flood protection purposes.

One of the key findings of the study was, "The net effect of the use of shore-parallel walls in estuaries can be significant reduction or elimination of sandy beach environments,"(Jackson, 1996). The study also concluded that the use of beach fill resulted in the most dramatic changes in physical characteristics of the shoreline areas including dune building and increased beach heights and widths. Also of interesting note, the study recognized that response to shoreline erosion is moving away from individual actions, to more geographically comprehensive, government-planned efforts.

TERMINOLOGY OF SHELTERED COASTS

The terms of reference of this study refer to the erosion of sheltered shore-lines. However, the meaning and perception of these terms— erosion, sheltered, and shoreline—varies among the interested parties and stakeholders including homeowners, coastal zone managers, geologists, engineers, and lawyers. In an attempt to address anticipated confusion, the meanings of these and other key physical resource features, as used in this study, are discussed below.

Shore

Definitions of the shore, including both open and sheltered coasts, range from the single location of intersection of sea level with the land profile (the *shoreline*) to inclusion of the entire region affected by wave action, spanning the inner continental shelf to the upper reaches of extreme storm swash (the *shore zone*). Often the definition appropriate for legal applications is quite different from the most obvious definition for understanding the physics. Figure 1-1 provides a useful illustration of the complexities of defining this common term.

Legal definitions of the shoreline are typically based on the location of a contour line at the elevation of mean high water or an equivalent tidal statistic. In a simple situation forced only by tides and waves on beaches with a rapid equilibrium response, this definition could be sensibly enforced. However, delineation of this boundary is complicated by the wide range of time scales over which sea level varies (from millennial glacial and interglacial climate periods to decadal climate fluctuations including El Niño to annual or seasonal storm surge signals) and by the range of time scales over which the beach responds to that forcing (interannual, annual, and single storm event). In addition, the value of

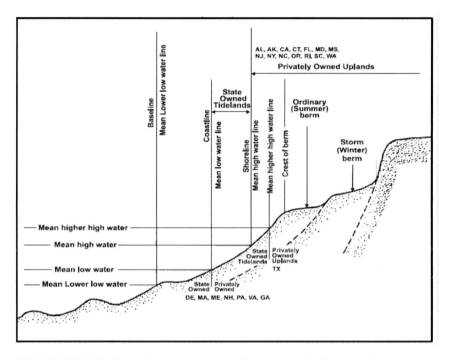

FIGURE 1-1 Definitions of nearshore zones and legal ownership for a typical coastal zone. The common legal demarcation between private and public lands, the "shoreline," is usually taken to be the intersection of the mean high water line with the beach profile. But temporal variation of the beach profile and even of sea level complicates this interpretation.
SOURCE: Modified from National Oceanic and Atmospheric Administration (NOAA), 2001.

any selected elevation statistic may be limited by the accuracy of estimates based on data availability. Local effects due to variability of wave-induced setup, local geostrophic or nonlinear shelf currents, variations in thermal expansion, among other effects, are neglected. Typically, a value is simply selected for regional application. For shores backed by sea cliffs or bluffs, the location most relevant to defining erosion may not be that of a specific elevation contour, but may instead be the location of the bluff top.

To describe the physics of coastal processes requires consideration of the "shore zone," defined as the active volume of sediment affected by wave action. However, this region can be difficult to bound, spanning from a poorly defined depth-of-no-motion on the seaward side, to some onshore boundary that might or might not include vegetated dunes and back-dune areas occasionally overtopped by extreme storm waves, or sea cliffs eroded by wave undercutting.

Erosion and Inundation

Erosion and inundation both result in the landward movement of the shore-line contour. These processes occur over a full range of time scales, including short-term events (waves, tides and storms) and chronic, long-term sea-level rise. Implicit in the definition of erosion is a choice of time scale, with longer events considered erosion and shorter events variance. Thus, a landward shoreline movement that recovers prior to the next storm would not usually be considered erosion, despite the potential loss of property associated with that variation.

Inundation refers to the temporary submergence of typically dry lands when there is an exceptional rise of the sea surface, and floodwaters cover the adjacent low-lying land.

Because shoreline is commonly based on movement of a single contour, it is precariously sensitive to details of the dynamics that determine how shore profiles adjust to natural forcing. This can introduce legal and operational difficulties. In the 1990s, The Netherlands determined a course of action to combat the ongoing threat of erosion of its shores, making it a legal requirement to mitigate erosion beyond the shoreline location defined in a national survey of 1990. Recognizing the sensitivity problems associated with a single contour definition, they instead defined a "momentary coastline" (MCL) based on the mean shoreline location integrated between −5 meters and +3 meters (approx. −16 feet and +10 feet) from NAP (Normaal Amsterdams Peil, or Amsterdam Ordnance Datum, which is the Dutch reference for sea level) (Figure 1-2). By basing the mitigation criterion on a shore zone definition, the law became appropriately robust to short-term fluctuations.

Sheltered Coasts

The term sheltered coast is frequently used to describe the shorelines of estu-aries and bays. These shores are considered sheltered because they typically abut smaller bodies of water with shallower depths and there is a limited distance over which waves can be generated, such that the waves are significantly less energetic than typical on an open coast with the same winds. However, there appears to be no objective criterion for the size of the body of water or the degree of energy reduction that would unequivocally distinguish a shoreline as sheltered.

The shoreline forms a continuum from open to sheltered areas—in some cases a physical feature may provide a clear demarcation, but in other areas there is a gradual transition from the open to the more sheltered environments. Compared to the typically long linear nature of open coasts, however, sheltered shorelines exhibit a more irregular configuration and often display very distinct geomorphic compartments that contain a complex of resources, the mix of which may change from compartment to compartment. This variation creates conditions that are both more varied and more complex than open coasts. Shores frequently included in the "sheltered" category range from low to medium wave energy.

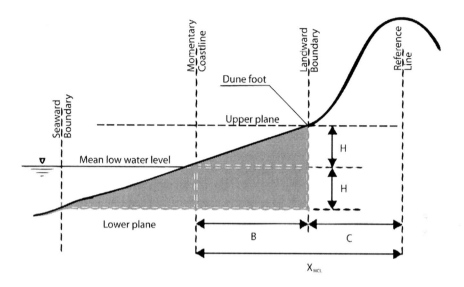

FIGURE 1-2 The definition of the open-coast shoreline used in The Netherlands is called the Momentary Coastline (MCL) and is found as the total area of beach sediment lying between −5 meters and +3 meters (approx. −16 feet and +10 feet) from NAP divided by 8m. This quantity integrates detailed profile fluctuations that would otherwise confuse estimation based on a more traditional shoreline location.

NOTE: H = Height between dune foot and mean low water [m]; A = Momentary Coastline Zone [m2]; B = A/2H = Momentary Coastline position [m]; C = Distance dune foot to reference [m]; XMCL = B+C = Momentary Coastline position + Distance dune foot to reference [m].

SOURCE: van Koningsveld and Mulderc, 2004. Courtesy of the Coastal Education and Research Foundation.

Instead of high energy waves, strong currents, storm events, or large boat wakes may be the primary drivers of sediment mobilization and transport on sheltered coasts. In lower energy environments, retention of smaller grained sediments creates mudflats and supports the growth of subtidal and intertidal vegetation. Overall, sheltered coasts describe a much greater diversity of conditions than found on open ocean coasts, requiring more site-specific approaches for managing erosion. Figure 1-3 illustrates the variety of shoreline configurations and environments found on sheltered coasts.

Many studies have focused on mitigating erosion on open ocean shores that are exposed to constant wave and current forcing. Thus, a focus of our early discussion was on defining those aspects of sheltered shores for which the physics and mitigation strategies were not simply scaled-down versions of their

FIGURE 1-3 Aerial image of Lower Machodoc Creek on the Potomac River in Westmoreland County, Virginia, showing the irregularity and diversity of shore types. The insets are six typical shoreline profiles around sheltered coasts.
SOURCE: Aerial image courtesy of the Commonwealth of Virginia.

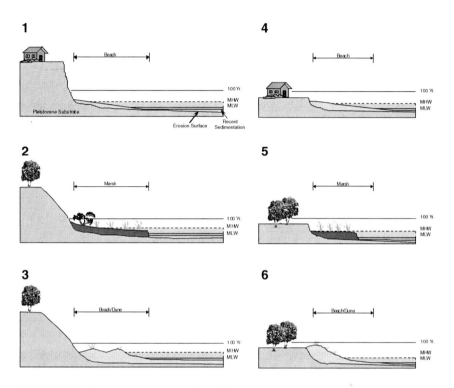

FIGURE 1-3 Continued. Insets as referred to on previous page.

open-ocean counterparts. Compared to their open-ocean counterparts, waves on sheltered shores are typically shorter, steeper and more episodic, with fluid energy often insufficient for sediment mobilization for extended periods of time. Thus, morphologies sculpted during storms often persist through subsequent calm periods as relict features. Local surge and currents generally play larger relative roles compared to waves in shaping sheltered shores.

Sheltered coasts consist of various combinations of geomorphic settings, or features, such as unconsolidated upland bluffs, dunes, beaches, intertidal (e.g., marsh, mangroves) and subtidal (e.g., macroalgae, seagrasses) vegetation, tidal flats, and sandbars. The definition of the three major geomorphic categories (beaches and dunes; bluffs; and mudflats and vegetated communities) are described below. The location and extent of any of these features is dependent on site-specific conditions. While some sheltered shorelines include hard rock outcrops, the erosion of features such as rock cliffs or shore platforms on sheltered coasts is considered a slow process and one unlikely to result in the need for the protective shoreline measures that are the focus of this study.

Beaches and Dunes

The shorelines most commonly used for recreational purposes (e.g., sun-bathing, walking, swimming, fishing) are those characterized by sandy beaches. Beaches are accumulations of any type of unconsolidated material that can be transported by wind or waves, from fine sand to cobbles. Beaches extend land-ward from the Mean Low Water Line to the place where there is a change in material or physiographic form, or to the line of permanent vegetation. In some cases, wind transported sediment will be deposited and accumulate along the backshore and form dunes.

The morphology of beaches and dunes is dependent on the interactions of the available sediment supply with the energy from waves, tides, and wind. Beaches are very dynamic systems. Storms have steep short waves that tend to lower and flatten beaches. Berms can be removed, the upper beach is eroded and dunes are cut as the beach adjusts to a lower volume of sand. The sand removed from the subaerial part of the system during storms accumulates just below low tide level to produce a wider flatter beach with a nearshore bar that allows storm wave energy to be dissipated. Under nonstorm conditions the fair-weather waves bring the sand in the bar back to the beach and wind actions restores the dunes. Such changes are frequently seasonal and result in a continually changing beach-dune form.

Bluffs

Coastal bluffs (also commonly referred to as banks) are elevated landforms composed of partially consolidated and unconsolidated sediments, typically sands, gravel and/or clays that are generally located landward of a beach or marsh.

As sea level rises, waves will attack the bluff face and toe of the slope, gradu-ally undercutting the bluff and causing sections of it to fall away. The erosion rate is dependent upon the degree of consolidation; rocky bluffs are not readily eroded by the forces that characterize sheltered coasts. Bluffs are also subject to erosion from other natural and anthropogenic events. Natural events include wind that strips vegetation and loosens sediment and groundwater seepage that undermines the vertical integrity of the structure; this is exacerbated in cold climates where freeze/thaw cycles speed disintegration of a cliff face. People contribute to bluff erosion when they build on top of bluffs and create places for water to enter and destabilize the system, they trample the vegetation that supports the steep cliff and sometimes mine the cliff for building material. Bluff erosion provides a sedi-ment source for other coastal features, such as the fronting beaches.

Mudflats and Vegetated Communities

Mudflats are intertidal areas with relatively fine sediment that can be veg-etated by plant communities (marshes or mangroves) or be barren in which case they are colonized by microscopic plant communities (microalgae) and bacteria. While marshes and mangroves are found in the intertidal area where they are regularly flooded during every high tide, seagrasses and seaweed occur in the subtidal area where they are submerged most of the time, except for extreme low tides when the plants in the shallowest areas may become exposed to air for a brief period of time. Temperature zone determines if marshes or mangroves will be found in the intertidal: Mangroves occur in warmer climates (tropical, warm temperate) while marshes are found in cooler climates (temperate). Substrate type determines if seagrasses or seaweed will be found in the subtidal: seagrasses dominate soft (sandy and muddy) substrates while seaweed dominate on hard (rock) substrates. These plant communities as well as the mudflats support a highly diverse and productive number of associated animals. Therefore, techniques to mitigate shoreline erosion that change the substrate characteristics will lead to changes in the associated plant and animal communities.

STUDY ORGANIZATION

The committee met three times during the course of the study. The first meeting, held in Washington, DC, in June 2005, provided the committee with an opportunity to discuss the background and study expectations with representatives from the sponsor agencies; Environmental Protection Agency, National Oceanic and Atmospheric Administration-Cooperative Institute for Coastal and Estuarine Environmental Technology, and the U.S. Army Corps of Engineers.

In addition, the committee developed plans for a workshop that was subsequently held in Seattle, WA, in October 2005. The purpose of the workshop was to provide the committee with additional background information, largely focused on an analysis of options available to mitigate erosion of sheltered coasts. In planning this activity, the committee decided not to limit the discussion to marine or estuarine areas, but to include experts from the Great Lakes. The rationale is that if the conditions leading to erosion are comparable in these bodies of water, then the issues arising with efforts to mitigate erosion will be similar. Additionally, the Great Lakes are recognized as subject to the Federal Coastal Zone Management Act (16 U.S.C. 1450 *et seq*) and with the exception of Illinois, the adjoining states have federally approved Coastal Management Programs containing many of the authorities and mechanisms in place to address the recommendations of this report. The workshop explored the geomorphic settings of sheltered coasts as well as how various erosion measures affect those settings. The workshop brought together approximately 32 professionals with such diverse expertise as: state and federal regulatory matters, science, engineering, land use planning, and legal issues. The participants came from around the continental United States and provided expertise on the range of erosion problems in various coastal regions. The complete agenda and participants list for the workshop are available in Appendix C.

The report is organized to address the issues outlined in the charge to the committee described above. Chapter 2 explores the physical processes of shoreline erosion and the erosion mitigation strategies used to address those processes. Techniques used to address shoreline erosion are discussed in Chapter 3. This chapter identifies four broad categories of options:

1. Land use regulation and management;
2. Vegetative stabilization;
3. Hardened structures (armoring the shoreline); and
4. Trapping or adding sediment.

Chapter 4 covers ecosystem services and values and how they are affected by shore erosion and mitigation measures, including living and nonliving components of the coastal habitats and the impacts of the most common structures installed to prevent erosion (revetments, breakwaters, seawalls, groins, and pilings). Chapter 5 describes the various regulatory, engineering, esthetic, and financial considerations that contribute to the decision-making process for mitigating erosion. This is followed in Chapter 6 by a description of a new shoreline management framework that synthesizes the committee's major findings and recommendations.

2

Understanding Erosion on Sheltered Shores

This chapter discusses the fundamental processes controlling erosion on sheltered shorelines, principally conservation of sediment mass and control of sediment fluxes. These basic physical laws provide the framework and organization for discussion of mitigation strategies. More complete discussions can be found in, for example, Krone (1962), Komar (1998a), or the Coastal Engineering Manual (USACE, 2000). These physical properties apply to both open and sheltered coasts. However, because of the segmented nature of sheltered coasts, the manifestations of these physical laws will vary in keeping with the variety of sheltered coasts habitats.

THE PHYSICS OF COASTAL EROSION

Inundation refers to the superelevation of sea level above a fixed topography. Short-term inundation (for example due to storm surge or heavy rains) is referred to as flooding, whereas longer-term (from a human perspective) coastal inundation results from sea-level rise. As discussed in Chapter 1, coastal erosion is often defined in terms of the movement of shore contours, and can be caused either by sea-level rise or by removal of geologic materials that make up the shoreline. Because rising sea level exposes portions of the shoreline to actions of waves and current, sea-level rise can exacerbate erosion. The principal tool for understanding erosion is the law of conservation of sediment mass, which requires information on the sediment (grain size, composition, sediment type) and transport capacity to estimate fluxes of sediment within the nearshore region.

Understanding Sediment

Geologists or soil scientists categorize sediment primarily on grain size, as sediment grain size controls many aspects of its behavior. Clastic sediment is the product of the breakdown of preexisting rock through physical or chemical weathering processes, which result in the vast majority of sediment on the Earth's surface. (The breakdown of shells or biological material, and physiochemical processes can also create sediment, for example the precipitation of aragonite needles from seawater can create calcium carbonate mud in some settings, but these sources are far less significant, volumetrically, than sediment derived from the breakdown of pre-existing rock). Individual sediment grains are classified on the basis of the physical dimensions of single particles. The 4 major grain size classes are gravel (> 2.0 mm [>0.08 in]), sand (62.5–62500 μm [0.0025–0.08 in]), silt (3.9–62.5 μm [0.00015–0.0025 in], and clay (< 3.9 μm [< 0.00015 in]). Similarly, a collection of grains, collectively referred to as sediment, is classified based on the distribution of grain size within a given volume (see Figure 2-1). The variety of sheltered coast types means that the spectrum of grain sizes and sediment classifications are reflected in sheltered coasts. For example, as described in Chapter 1, mudflats tend to consist of relatively fine sediments such as clays; and bluffs are typically a mix of sand, gravel, and clay.

Size is a fundamental predictor of the ability of water to entrain sedimentary grains from a streambed or to transport grains by current or wave activity. Con-

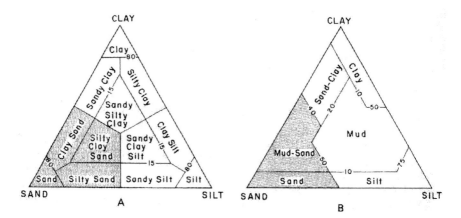

FIGURE 2-1 Two examples of nomenclature for classifying sediment types; this classification is for mixed clastic sediments formed from varying contributions from sand, silt, and clay fractions: (A) symmetrical conceptual scheme and (B) asymmetrical scheme, the latter based on actual usage of marine geologists. The numbers indicate the percentage of each component in the total composition along that axis.
SOURCE: Pettijohn et al., 1973. Courtesy of Springer-Verlag, New York.

versely, the current or wave energy that characterizes various settings (referred to as the hydrodynamic setting) determines to a large degree the grain size of sediment deposited along the bottoms of streams, bays, estuaries, or other water bodies. Large waves can mobilize and transport larger grains than can small waves, thus the ability of wind-driven waves to entrain or transport sediment of various sizes may change through time as winds change in intensity or direction. Similarly, physical features that focus wave energy can alter how waves generated by a more or less uniform wind velocity impact various portions of shoreline. Sediment of various mineral composition or grain sizes also exhibit different mechanical properties and various biological organisms interact with different sediments in different ways. Thus understanding the linkage between sediment dynamics and hydrodynamic setting is a key aspect of predicting and hopefully controlling the erosion of coastlines.

Conservation of Sediment Volume

The equation for conservation of sediment volume[1] can be written as follows:

$$\frac{\partial h}{\partial t} = \frac{1}{\alpha}\left[\frac{\partial Q_x}{\partial x} + \frac{\partial Q_y}{\partial y}\right] + Q_s \qquad (2\text{-}1)$$

In words, Q_x and Q_y are the sediment fluxes in m^3 per m width of flow per second and represent fluxes in the cross-shore (x) and longshore (y) directions, respectively. $\partial h/\partial t$ is the rate of change of water depth, that is a function of erosion or accretion. α is a packing coefficient that accounts for the fact that sediment does not pack perfectly as they settle, but will leave gaps that are filled with pore water. A key point in Equation 2-1 is that erosion is not caused by sediment transport per se, but only by gradients in transport. Thus, a steady, uniform transport of sand will not affect the beach profile. However, if gradients in transport exist (e.g., $\partial Q_y \partial y$ is not equal to zero), either more (negative values) or less (positive values) sediment is entering the area than is leaving. That surplus or deficit is reflected in increasing or decreasing accumulations of sediment, hence a time rate of change of depth, $\partial h/\partial t$. More simply stated, divergences of transport (positive gradients) cause erosion, while convergences (negative gradients) cause deposition and accretion (offshore migration of the shoreline).

In Equation 2-1, α represents the packing efficiency of sediment in the bed. Q_s represents sources or sinks of sediment in the system, perhaps inputs from rivers or even biogenic production of sediments, or losses due to breakdown of

---

[1]Sediment mass and volume are uniquely related by the sediment density and a packing factor, α, representing the fraction of sediment in a bed that is sediment versus pore water. For simplicity, the terms sediment mass and sediment volume will be used interchangeably in this report.

nonresistant material. Overall, the key to understanding or controlling erosion is to understand or control sediment transport or sediment fluxes.

Sediment transport can be considered in terms of three processes: (1) initiation of sediment motion (sediment threshold conditions, also referred to as entrainment), (2) processes and magnitudes of sediment movement or flux, and (3) processes of sediment settling back to the bed (also referred to as deposition). Mitigation strategies attempt to leverage one or more of these processes.

Threshold of Sediment Motion

Settled sediments do not move in simple proportion to the strength of waves and currents. Instead, initial grain motion requires the flow to exceed some threshold that depends on the size and density of the grains. The details of this relationship are complex and largely empirical, but the general concept compares the size of the fluid bottom stress, τ (trying to shear a grain from the bed) to the immersed weight of the grain. The ratio of these two forces is used to define the nondimensional Shields' parameter, θ, (Shields, 1936),

$$\theta = \frac{\tau}{\rho(s-1)gD} \tag{2-2}$$

where ρ the density of water, s the relative density of the sediment compared to water, g the acceleration due to gravity and D the diameter of the sediment grains. For unconsolidated and noncohesive sediments, the value of the Shields parameter, θ_c, above which sediment is mobilized, can be readily found (see, for example, Nielsen, [1992]). Mitigation strategies often involve increasing the denominator of (2-2) through the use of large or dense protection material that will not be moved under expected wave and current stresses.

Fine sediments such as muds and silts can be cohesive, with electrostatic or chemical bonds between adjacent grains that make them harder to dislodge than the free, independent grains discussed above. Thus, values of θ_c are typically much larger for cohesive material and the sediments are much more stable than would be expected, given their small grain sizes (see Figure 2-2). Cohesion is increased in salt water over fresh and can also be fostered by biological processes that may bind sediments together, thus increasing their resistance to transport.

Shorelines that respond to wind wave energy span a range from mudflats[2] to rocky shores. These shorelines respond to incident wave energy through adjustments in planform and profile. The threshold for sediment movement is one of the main differences between sheltered and open coasts. Sheltered shores, because they characteristically experience lower wave energies and slower moving currents, commonly are composed of finer sediments falling below the threshold for movement except during extreme events such as storms. Also, because of the

---

[2]Vegetated mudflats are commonly referred to as marshes (see Chapter 1).

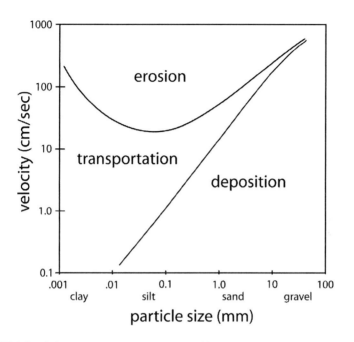

FIGURE 2-2 Hjulstrom Diagram showing the differing behavior of sediment particles of various size in unidirectional (current) flow of varying velocities. The behavior is characterized by three fields, no transport or deposition (particles settle out of water column), transport (dark gray band central band where grains are actively transported in the water column), and erosion (where grains are entrained into the water column). Note that fine-grained clay and silt-size sediment, with its greater grain-to-grain cohesion, requires more energy to be entrained than does fine sand. Although more complicated, the behavior of these sediments in oscillatory (wave) flow is somewhat analogous in that deposition, transport, and erosion of a specific sediment type requires increasing flow velocities.
SOURCE: Derived from Hjulstrom, 1939.

variety of conditions on sheltered coasts, responses vary depending on the type of coastline; for example, depositional processes on mudflats are largely dependent on tidal processes. The role of waves and tides in controlling mudflat morphology is complicated by the fact that the energy input of neither the waves nor the tide is independent of the morphology (Pethick, 1996). Depositional processes are usually considered to be dependent on the suspended sediment concentration, the sediment fall velocity, and a subcritical bed shear stress. In contrast, erosion depends on bed sediment density and occurs when bed shear exceeds the threshold stress (Krone, 1962). More recent field studies have examined the terms "sedimentation" and "erodability" in more specific terms (Amos et al.,

1998). The response of mudflats to short-term changes in wave energy differs from sandy systems as the cohesive properties of the deposited sediment result in erosion threshold being greater than deposition thresholds. This difference is amplified by seasonal effects of biological activity that can act to bind or disturb surface sediments (e.g., Ruddy et al., 1998).

Equation 2-2 represents a balance between the fluid stress acting on the grain to move it from the bed and the particle weight acting to keep it in place. This relationship can be complicated if pore water infiltrates into or exfiltrates from the sediment bed. The role of these cross-bed fluxes on the resulting threshold criteria is not totally clear, but has been examined by Turner and Masselink (1998) among others. Some mitigation strategies have been based on lowering of beach groundwater levels to encourage infusion across the sea bed, thereby stabilizing beach face sediments.

Sediment Transport Processes

Once freed from the bed, sediment is carried by the combined action of waves and mean currents. The physics of this sediment transport are complicated and the subject of many books (e.g., Nielsen, 1992; Fredsøe and Deigaard, 1992). Transport requires two elements: sufficient fluid turbulence or energy to maintain sediment motion against the tendency to settle and a mean flow to cause net transport (or a nonzero time mean term for oscillatory flows).

The resulting net transport is directional and can be resolved into cross-shore and longshore components. The longshore component of transport is most easily understood since it can readily be thought of as the result of waves stirring sediment into suspension then subsequent transportation by a steady current along the coast (Komar and Inman, 1970). Such a longshore current most commonly occurs when waves approach the shore obliquely and drive a current along the shore in the surf zone (e.g., Bowen, 1969). Convergences or divergences of this transport, for example on the updrift and downdrift sides of a jetty, cause accretion or erosion, respectively (Equation 2-1). This is one of the primary mechanisms of shoreline erosion.

Cross-shore transport processes are not as simple since any steady onshore current will be blocked by the shoreface. However, undertow plays a role, as do asymmetries in the shapes of waves as they become nonlinear in the surf zone (e.g., Bowen, 1980; Bailard and Inman, 1981). On open coasts, the beach profiles observed in nature represent a dynamic balance between offshore and onshore transport processes, with offshore dominating during storms and onshore in intervening calms. This onshore-offshore movement of sediment within an active profile envelope appears to cause changing morphologies including sand bars, but does not usually seem to cause significant net loss of sediment (Bowen, 1980). For sheltered coasts, episodic wave energy and the common occurrence of sub-threshold wave orbital velocities can cause stranding of storm deposits offshore.

SPATIALLY AND TEMPORALLY VARIABLE FACTORS CONTROLLING COASTAL EROSION

The fundamental physical processes described above operate universally. However, regional variability in the specific conditions determines which processes will dominate the nature of any particular portion of a coastline. On open, high energy coastlines, fine-grained sediment is rapidly winnowed out leaving only sand size or coarser material. Thus, the typical beach of open-ocean shorelines is obviously different, even to the layman, than the majority of shorelines along sheltered coasts. Thus, in order to apply an understanding of basic physics of sedimentary systems to sheltered coasts, one needs to understand how regional variability influences the character of sheltered coasts for a number of parameters and geographic scales. This section discusses how the unique characteristics of sheltered coasts interact with the factors that control coastal erosion.

Wave Climate

The primary source of energy for the suspension and transport of sediment on most sheltered coasts is associated with surface waves, generated by the wind. The maximum size to which these waves can grow depends on the amount of energy transferred from the wind. This, in turn, is a function of the strength of the wind, the duration for which it blows, and the span of water over which the waves can grow before leaving the forcing area or run into a shore (this distance is called the fetch). For sheltered shores, wave growth is, by definition, limited by the fetch. Waves are characterized by their period, the time between passage of adjacent crests, and their height (vertical extent). The height attained by fetch-limited waves depends on the square root of the fetch, while the energy carried by the waves varies linearly with fetch. Thus, the size of the body of water adjacent to a shoreline, particularly the distance in the direction of strong winds, has a strong influence on the potential severity of damage due to the worst-case storm waves. Because sheltered coasts are fetch limited, the impact of wave height and energy on erosion depends on the wave energy generated during storm events as constrained by the fetch in the direction of the storm. Fetch limitation affects the design of erosion mitigation structures by placing a cap on the possible energy that a structure will face.

Extensive discussion of wave generation by winds, including equations for wave heights, periods and spectral characteristics, is contained in many references, for example the Coastal Engineering Manual (USACE, 2000).

The previous discussion deals only with wind-generated waves. The safer waters of sheltered areas often attract a variety of both recreational and commercial boat traffic whose wakes may have an erosive impact larger than the wind-generated waves. In contrast to wind-generated waves, there are no simple equations to predict the magnitude or impact of boat-generated waves. Thus, site characterization of potential boat-wake effects must be conducted on a case-by-case basis.

Sediment Type

Sheltered coast sediments represent the spectrum of grain sizes and sediment types, depending on the geomorphic setting. Shorelines along sheltered coasts may be made up of sediment which has undergone various degrees of consolidation, ranging from recently deposited unconsolidated sediment to partially cemented rock (characterized by its low strength) or, in some settings, exposed bedrock. Bedrock generally represents the antecedent geology of a given location and thus may be made up of exposed sedimentary, metamorphic, or igneous rock that crops out along shoreline. Depending on the climate of a given region and the nature and attitude of the exposed bedrock, these outcrops may form various types of resistant rock headland, including steep cliffs or shallow ramps along the water's edge. Because sheltered coasts, by definition, lie along water bodies with limited fetch, the current and wave energy that characterize these water bodies generally is insufficient to rapidly erode fully indurate bedrock, thus such settings are not explored further in this chapter in the remainder of the report. Rather the report will focus more exclusively on shorelines made up of various types of sediment.

As defined in Chapter 1, the term "beach" is used generally to describe gently sloping shorelines characterized by accumulations of sand to cobble-sized sediment, while "mudflat" refers to a gently sloping shoreline characterized by clay and silt-sized sediment. Beaches and marshes are often backed by coastal bluffs (also commonly referred to as banks), elevated landforms composed of unconsolidated sediments, typically sands, gravel, and/or clays.

Unconsolidated sediment will maintain a slope at a given angle with respect to the horizontal (referred to as the angle of repose) depending on grain size. Because fine-grained sediment provides a greater opportunity for grain-to-grain contact, friction between particles will allow the slope of a bank made up of unconsolidated clay- and silt-size particles to be steeper than that of a bank made up of sand-size particles. Furthermore, the stability of partially consolidated sediment of a given grain size is increased as compaction increases the number of grain-to-grain interactions, as roots or other biological material binds sediment together, and over a much longer time period precipitates from pore water supersaturated in certain minerals will cement the grains of sediment together.

The processes of bluff erosion thus differ from those governing beach or mudflat erosion. Bluffs erode due to slope stability failures, when the cohesive strength of the material in the steeply sloping "bluff face" is exceeded by the down-slope component of the weight of material being supported. Thus, both steepness and material strength are factors in bluff erosion and are the foci of mitigation strategies.

Strategies to address these two major factors in bluff erosion differ. Wave processes play a role in undercutting the bluff and removing collapsed material that temporarily protects the base of an eroding bluff, thus allowing over-steepening of the bluff face. Toe protection may be used to guard against this problem. The strength of the cliff material is most strongly affected by ground-water, and groundwater seepage from rain or irrigation can substantially reduce stability by either forcing sediment grains apart (which occurs when pore pressure exceeds lithostatic pressure due to the weight of the overburden) or by facilitating slippage along discontinuities in the sediment pile (for example water flow along clay layers). Thus in most settings where bluff erosion is a concern, groundwater control is a primary mitigation approach. For buildings constructed on a bluff, the design needs to include evaluation of the bluff conditions so that the additional overburden, modified water table, and construction activities do not destabilize the bluff.

Sources and Sinks

The final term in Equation 2-1, Q_s, represents a host of possible sources and sinks of sediment to the nearshore zone. Most obvious is river input, a source that is usually quantifiable but, for many cases, has been significantly reduced over the 20th century due to the damming of rivers. Sediments can be lost offshore due to exceptional storm events, discharge from large rip currents, and loss to nearby deep channels and submarine canyons, but sheltered coasts are mostly subject to sediment loss due to major storm events. Sea-level rise can be thought of as a sink of sediment equal to the active area of the shoreface times the rate of sea-level change (Bruun, 1962).

Bluff erosion is generally viewed as a source of beach sediments despite being a loss of property. Similarly, the overtopping of dunes and associated transport of sediment into the back-dune region is considered a sink of sedi-ments to the nearshore system despite being a source (albeit inconvenient) of sediment to the upland property owner. Sand mining from beaches is an example of an anthropogenic sink, while beach nourishment is clearly an anthropogenic source.

Littoral Cells

The principle of conservation of mass embodied in Equation 2-1 is a simple, yet very powerful tool for understanding coastal erosion and for defining littoral cells. Sediments can either enter (from sources) or leave the system (via sinks), but otherwise are simply redistributed through sediment transport. Erosion occurs at sites of divergence, areas where the amount of sediment mobilized and lost exceeds the amount deposited. Accretion occurs at sites of convergence, where sediment deposition exceeds sediment loss. If one can identify boundaries across which no transport occurs, one can define simple sediment budgets over inter-boundary regions.

Littoral cells are defined as sections of coast for which sediment transport processes can be isolated from the adjacent coast. Typical boundaries are large headlands, ends of islands or estuaries that either stop transport or act as quantifiable sinks. Within each littoral cell, a sediment budget can be defined that describes sinks, sources and internal fluxes (sediment transport).

The concept of littoral cells emphasizes the interconnectedness of nearshore systems. For example, an interruption of longshore transport within a cell by a groin will result in a sediment transport convergence, hence accretion at the local property, but an equivalent divergence and erosion will occur on the downdrift side. In fact, any action taken within a cell will affect the surrounding coast. Ideally, the interconnected nature of the system should be understood and accounted for when mitigation strategies are considered. This concept is particularly useful for addressing erosion on sheltered coasts, which tend to be highly compartmentalized with littoral cells on smaller scales than open coasts. However, permitting for erosion mitigation is commonly done on a lot-by-lot basis without consideration of the regional implications. It is critical to recognize that nearshore processes occur at regional scales and that littoral cell processes in particular represent the appropriate physical process unit for planning mitigation strategies.

The Role of Changes in Sea Level

The previous discussion only deals with half of the problem, land loss due to the removal of sediments from the underlying shore profile. Landward shoreline movement will also occur for the case of a fixed shore profile on which there is a long-term rise in sea level relative to the land. Shoreline movement associated with sea level may change on time scales of decades or longer and is often considered to be an erosion problem, even though it does not necessarily involve the removal and transport of sediment.

There are many causes of long-term sea-level change, each occurring on different temporal and spatial scales. At the largest scale, eustatic or global sea level is rising due to melting of the polar ice pack and thermal expansion of seawater. Eustatic sea levels have risen between 100 and 250 mm (about 4 and 10 inches) during the past century and will inevitably be affected by climate change in the future. The rate of eustatic sea-level rise during the twentieth century has been nearly 2 mm (0.1 inches) per year, which is an order of magnitude higher than the average over the last several millennia. By 2100 the projected rise is 90 mm to 880 mm (3.5 to 34.5 inches; IPCC, 2001). On passive margin coasts (coasts that are not tectonically active) with slopes between 1:100 and 1:1000, a 90 mm (3.6 inches) eustatic rise would result in a corresponding landward shift of the shoreline of between 90 and 900 meters (about 280 and 3000 feet). In addition to global increases in sea level, local subsidence due to sediment compaction or fluid withdrawal, or regional subsidence due to isostatic adjustment of the seafloor due to sediment or tectonic loading (e.g., the vast accumulation of sediment that make up the Mississippi and Bengal deltas actually compress the oceanic crust beneath them) or density increases of the lithosphere due to cooling of igneous intrusions (e.g., seamounts cool and subside through time as they move away from the underlying source of volcanism) can contribute to the relative sea-level rise experienced at a given location along the coastline. As a consequence, projections of eustatic sea-level rise may significantly under predict the landward shift of the shoreline in areas which experience high rates of relative sea-level rise such as coastal Louisiana, where the mean sea-level trend from 1947-1999 (at Grand Isle) was 9.85 mm/year (about 0.4 inches) (NOAA, 2006). Also, see NRC [2006] for more discussion of the factors responsible for coastal erosion in that area. Regional uplift due to large-scale glacial rebound from the last ice age, and slow uplift along convergent margin coasts like those of the U.S Pacific Northwest (Figure 2-3) can partially or fully offset eustatic sea-level rise. Figure 2-4 illustrates the regional variation in sea-level trends around the United States over the past century.

The nature of the nearshore system response to rising sea level for both roll-over and hold-the-line strategies is an ever-increasing stress on the system followed by failure and roll-back. However, for the hold-the-line approach, longer periods of stability are traded for greater eventual catastrophe. In both cases, the system follows the laws of self-organized criticality (see Box 2-1), rather than of equilibrium response to random stimuli.

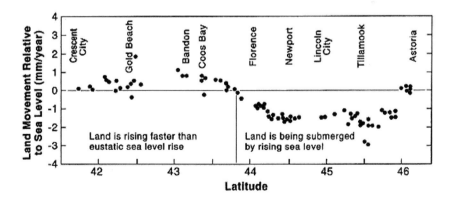

FIGURE 2-3 Map of along-coast variations in relative sea-level rise from Northern California to Oregon found from long-term leveling data. Variations are due to tectonics associated with convergence of the offshore ocean plates with the North American plate. NOTE: 1 mm is approximately 0.04 inches.
SOURCE: Komar, 1998b.

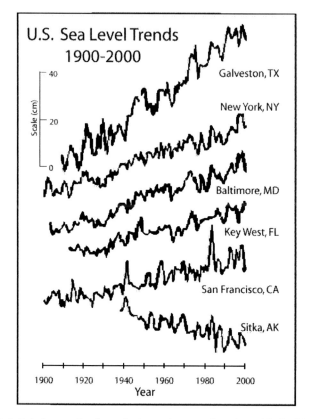

FIGURE 2-4 Relative sea-level trends, in centimeters, for selected U.S. cities from 1900 to 2000.
SOURCE: Data from National Oceanic and Atmospheric Administration; graph available through the U.S. Environmental Protection Agency, 2000.

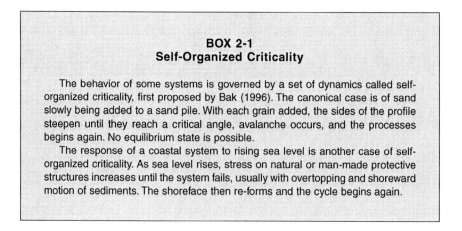

BOX 2-1
Self-Organized Criticality

The behavior of some systems is governed by a set of dynamics called self-organized criticality, first proposed by Bak (1996). The canonical case is of sand slowly being added to a sand pile. With each grain added, the sides of the profile steepen until they reach a critical angle, avalanche occurs, and the processes begins again. No equilibrium state is possible.

The response of a coastal system to rising sea level is another case of self-organized criticality. As sea level rises, stress on natural or man-made protective structures increases until the system fails, usually with overtopping and shoreward motion of sediments. The shoreface then re-forms and the cycle begins again.

IMPLICATIONS OF GEOMORPHIC SETTING
FOR EROSION MITIGATION STRATEGIES

As discussed earlier in this chapter, macroscopic features of shorelines vary substantially. The presence of bluffs or resistant rock headlands, the abundance and nature of local sediments, and the proximity of deep water, are characteristics of the local or regional geology that strongly affect local dynamics. Collectively these are known as the geomorphic setting of the coast. As described in Chapter 1, this report groups the geomorphic settings under three categories: (1) beaches and dunes, (2) mudflats and vegetated intertidal communities, and (3) bluffs. Different erosion mitigation approaches have been developed to address erosion in these three geomorphic settings.

A well-designed erosion mitigation project works in harmony with the local and regional geomorphic setting. Strategies for the mitigation of erosion on both sheltered and open shores may be thought of as attempts to influence one of the terms in Equation 2-1 describing the conservation of sediment mass. These approaches either attack the components of sediment transport or manipulate the source and sink term. Specific approaches to addressing erosion are described in Chapter 3.

Understanding the Physical Setting

Clearly, a number of interrelated factors play a role in the stability of a given portion of the shoreline. It is important to understand where a specific shoreline of concern lies within the larger sedimentary system (i.e., within a given shore reach, littoral cell or other geomorphic coastal unit). Understanding the geomorphic evolution is important for several reasons: (1) documenting, quantifying and illustrating shore change, (2) evaluating shore impact by natural and man-made features and (3) assessing nearshore changes in channels, shoals and sand bars. There are a number of open sources for this information. Historical aerial photography can be informative for understanding and documenting historical shore change (see Figure 2-5) and can often be found in planning offices and at the Natural Resource and Conservation Service (NRCS). Other sources of shoreline change data include reports, topographic maps, and shoreline change reports from the U.S. Geological Survey and historical charts from the National Ocean Service, NOAA.

Sheltered coasts (defined in Chapter 1) are found in estuaries, lagoons, or sounds where the shore exposure is fetch limited. Rising sea levels inundate old fluvial systems and create geomorphic units that are generally shorter (measured parallel to the shoreline) than units on open-ocean coasts. Sheltered shore segments, often referred to as reaches, might be 100 meters to a few kilometers in length (approx. 300 feet to a couple of miles) where each segment is bounded by tidal creeks, inlets, or abrupt changes in shore orientation. On open-ocean coasts,

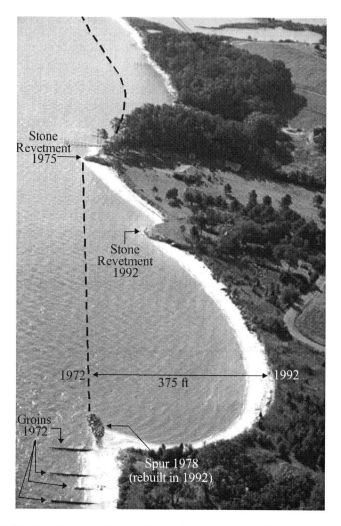

FIGURE 2-5 Shoreline evolution in Northumberland County, Virginia, where the width was 114 meters (approx. 375 feet).
SOURCE: Hardaway and Byrne, 1999. Courtesy of the Virginia Institute of Marine Science (VIMS).

reaches or littoral cells are often much longer, even hundreds of kilometers (or miles) in length.

A littoral cell consists of three broad zones: the zone of erosion, the zone of transport, and the zone of accumulation (or deposition; see Figure 2-6). No clear demarcation exists between these zones; rather, they gradually merge into one another in broad bands along the shore (Taggert and Swartz, 1988; Myers,

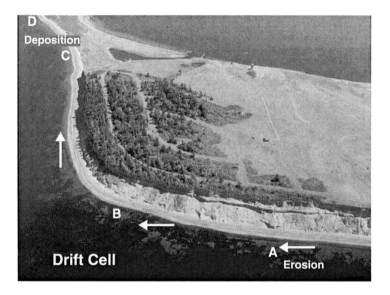

FIGURE 2-6 Shoreline of a littoral (or drift) cell that illustrates the transition from the zone of erosion (A), to the zone of transport (B), zone of deposition (C), and the less active terminus (D). The arrows indicate the direction of sediment transport. The change in zones reflects the transition from a wave-dominated erosional environment to a subaerially-dominated environment with increasing beach width. The location is in Puget Sound.
SOURCE: Photo modified from Myers (2005).

2005). A reach is a segment of shoreline that generally coincides with the lateral extent of a single littoral cell.

The rate of change is an important component of assessing the site before choosing a mitigation strategy. For example, Figure 2-7 shows bluffs (or banks) that are eroding at different rates, determined by the fetch exposure which would require different approaches to manage the problems caused by erosion.

Coastal segments, reaches, or compartments representing a single littoral cell (Figure 2-6) typically have three components, an erosive segment on one end, an accretionary segment on the other end (an area of active deposition), and in between, a transitional or transport segment that carries material from the erosive segment to the accretionary segment. This characterizes the shore planform of a reach. The coastal profile can be defined as a cross-section of transect across any shore segment that defines the upland, shoreline, and nearshore region as illustrated in Figure 2-8. This, in combination with the shore planform, provides a three-dimensional perspective of shoreline. Surveys of upland topography, the shoreline, and nearshore bathymetry are used to develop these profiles.

FIGURE 2-7 Three levels of bank erosion: Stable, Intermediate, and Unstable.
SOURCE: Hardaway et al., 1992. Courtesy of VIMS.

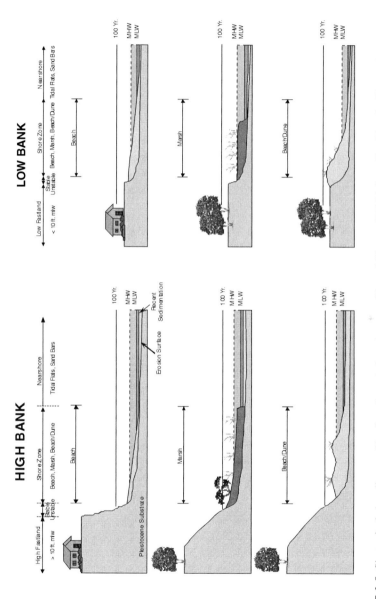

FIGURE 2-8 Six typical shoreline profiles found on sheltered coasts. Note: 100-year FEMA base flood level will vary by coastal locality. For example in Cape Charles, Virginia, the 100-yr flood level = 2.8 meters (approx. 9.3 feet) MLW (adjusted; USDHUD, 1992).
NOTE: Three meters is approximately 10 feet.
SOURCE: Hardaway and Byrne, 1999. Courtesy of VIMS.

Human Elements of Shoreline Character

Coastal development with shoreline hardening can reduce or stop erosion. Impacts of erosion protection on a few lots within a reach are generally not significant. However, the cumulative effects of shore hardening along major portions of a reach can alter the shore zone, causing loss of beach, reduced sand supply and transport, and a deeper nearshore region. When erosion is addressed on a site-by-site basis, the solution for one site can have the effect of creating an erosion problem on the neighboring property. Since property owners only address erosion on their shorelines, it is important that the contractor, the consultant, and local permitting authority all understand the potential cumulative impacts. A shoreline management plan that considers the impacts of continued modification of the shoreline can prevent the unintentional domino effect of armoring which may eliminate the natural shoreline along the reach.

Erosion control projects can also have significant impacts on the uplands. For instance, grading of shoreline banks for construction might include removal of a forested area that helped stabilize the slope which should be revegetated after construction of the new bank. In addition to stabilizing the slope, upland vegetation, particularly trees and shrubs, filter nutrient-laden water runoff and reduce loads entering coastal waters. However, trees should not be planted too close to the edge in coastal areas with marsh fringes or seagrass beds, because the leaf canopy may excessively shade the marsh and seagrasses in the intertidal and subtidal zones and stunt the growth of the grasses. Some shore protection methods have been employed that do not include bank grading, but instead allow the bank face to continue to erode to an equilibrium profile.

Finally, anthropogenic impacts include not only shoreline structures but also nearby dredging projects and piers. Although the latter two are not the focus of this report, they often affect coastal processes and consequently influence the shoreline. The number of pier installations has increased on many sheltered coasts, a function of coastal development and higher demand for access to nearshore waters.

FINDINGS

Understanding the interplay of sediment composition with the hydrodynamic setting of a given shore reach is critical in managing shorelines along sheltered coasts. Designing and implementing an effective mitigation strategy requires understanding how altering one or more of these sedimentation and hydrodynamic factors discussed in this chapter will affect the overall system. Each of these four options should be considered:

• What will happen if nothing is done? The erosion processes will most likely continue into the near future unless there is historical evidence of an impending change in patterns of shore evolution.

- What if vegetation is planted on the shoreline?
- What will be the consequences of hardening the shoreline?
- What if sand is added (beach nourishment) or trapped (groins or breakwaters)?

The effectiveness of these four options in addressing erosion will depend on the context of the littoral system being evaluated. This topic is discussed in more detail in Chapter 3.

3

Methods for Addressing Erosion

Numerous techniques,[1] technologies,[2] and planning measures[3] are available to address the issue of shoreline erosion, with most methods primarily intended to protect property from shore erosion caused by wave attack. Other erosive forces at the regional and local scale may affect the site's geology and geomorphology, as described in Chapter 2, and some methods are specific to these forces. This chapter provides an overview of techniques and technologies commonly used to address erosion, followed by a discussion of important design elements and criteria that should be considered in selecting an approach to address erosion on sheltered coastlines.

APPROACHES TO EROSION

Techniques used to address erosion along sheltered coasts may be placed into broad categories, such as those proposed by Nordstrom (1992), Rogers and Skrabal (2001), and, more recently, Rogers (2005). Most guidelines and reports on shore protection employ the same basic concepts to discuss approaches such as structural or "hard" methods versus nonstructural or "soft" approaches (Hardaway and Byrne, 1999; Maryland Department of Natural Resources, 1992; New York Sea Grant, 1984; Pile Buck, 1990; Rogers and Skrabal, 2001; USACE, 1981, 1984; Virginia Marine Resource Commission, 1989; Ward et al., 1989; Eurosion, 2004).

---

[1]"Techniques" refers to broad categories of approaches used to address erosion.

[2]"Technologies" refers to specifically designed or engineered methods used to address erosion.

[3]"Measures" refers to regulatory and planning actions used to address erosion.

For the purposes of this study, four categories of commonly used techniques to address erosion are identified: (1) Manage land use, (2) Vegetate, (3) Harden, and (4) Trap and/or add sand. Each of these techniques has one or more specific type of technology or measure that can be used to meet its objective, discussed in the following sections. It is common for some combination of techniques to be applied at any particular location of a sheltered coast. For instance, if a decision is made to vegetate a site with a fringe marsh on a low to moderate wave energy coast, a combination of marsh plantings (vegetate) on sand fill (add sand), protected by a stone sill (harden) might be installed as a system. Although these techniques are discussed as separate topics, it is common for multiple methods to be used in combination.

Manage Land Use

Decisions on land use typically occur at the state and local levels. Land use measures have both spatial and temporal components. Spatial scales vary at the federal, state, regional and local levels. Historically, land use controls have been applied at the level of an individual lot without consideration of the system-level (e.g., littoral cell) processes that drive erosion. The temporal component derives from the requirement that the effectiveness of these measures depends on the consistency and longevity with which they are applied.

Management of land use varies greatly, from passive to active approaches. Measures to manage land use may be outlined as follows:

(1) planning
- managed retreat
- community visioning
- green planning
- education
- technical assistance
- restoration and reclamation
(2) regulation
- buffers
- setbacks
- down-zoning
- construction standards
- perpendicular access
- institutional reorganization and coordination
(3) incentives
- current use tax
- transfer of development rights
- conservation easements
- rolling easements

(4) acquisition
   • fee simple
   • conservation easements
   • rolling easement
   • lot retirement

Land use control and land management techniques transfer responsibility of shoreline management from the individual to the community and are often perceived as more difficult to implement than a single action by a property owner. The long-term individual and cumulative benefits of these measures extend beyond those produced by other methods, including: (1) reduced coastal infrastructure and development, (2) diminished water quality degradation, (3) improved ecological status of shorelands by avoidance of fragmentation, (4) no loss of recreational access, (5) increased property values, and (6) reduced property losses.

Vegetate

Vegetation can be used to control shore erosion by planting appropriate grasses into the existing tidal and supratidal substrate. This strategy is generally limited to sites with very limited fetch. At sites with a larger fetch (over roughly 0.8 km, about 0.5 mi), creation of a marsh fringe will require the addition of elements such as sand fill (to provide a better substrate or planting terrace, see Figure 3-1) with or without some type of sill to attenuate wave action (see Figure 3-9).

This procedure for addressing erosion is not limited to the shore zone, but can be used elsewhere, such as on upland banks or bluffs. Various forms of bioengineering techniques can be employed to control groundwater seepage and surface runoff. Vegetation also can be used to stabilize banks or bluffs—roots from plants (trees, bushes, grasses) bind soils and form a living, adaptive barrier. Vegetation can be used in combination with graded banks to provide an effective approach to reduce erosion.

Marshes

Marsh creation for shore erosion control can be accomplished by planting the appropriate species, typically grasses, sedges, or rushes, in the existing substrate and addressing the original cause(s) of marsh loss (e.g., altered hydrology, low water clarity, invasive species, erosion from boat wakes, or shading from overhanging tree branches on the bank). Planting of marsh grass to stabilize the shoreline has been used successfully for many years (Knutson and Woodhouse, 1983). Numerous planting guidelines exist for creating marsh fringes such as Rogers and Skrabal (2001). Recently, educational efforts by NOAA and others in Chesapeake Bay and North Carolina have resulted in a revival of these tech-

FIGURE 3-1 Preproject shoreline on Wye Island in Queen Anne's County, Maryland (top). Marsh grass was planted on sand fill and short, stone groins were placed (middle, 3 months after installation). Bottom is six years after installation.
SOURCE: Hardaway and Byrne, 1999. Courtesy of the Virginia Institute of Marine Science (VIMS).

niques. The term "Living Shoreline" has been coined to help promote interest in this method rather than using techniques that harden the shore. In Chesapeake Bay, particularly in Maryland, over 300 marsh fringe sites have been constructed, planted with marsh grasses, and observed for 15-20 years, with an impressive record of performance for erosion control and wetland habitat creation (Maryland Department of Natural Resources, 2006).

Seagrasses

Submerged vegetation such as seagrass stabilizes the sediment and may contribute to wave attenuation at low tide (Koch, 2001). The value of seagrass beds for shore protection is limited by their seasonality. During the winter months, seagrasses in temperate areas become less dense or may even disappear, providing less protection during the season when increased storm activity may bring increased wave activity. The highest degree of wave attenuation, and hence potential shore protection, occurs when seagrass occupies the full height of the water column (Fonseca and Cahalan, 1992). Water levels tend to be higher than normal during storm events and the capacity of seagrasses to attenuate waves (and provide shore protection) is diminished.

Replanting of submerged aquatic vegetation (SAV) is typically undertaken to restore habitat after these plants have been lost in the subtidal area. Planting techniques, including wave-exposure requirements, can be found in Fonseca et al. (1998). Light availability (at least 10 percent of surface irradiance) is essential for the long-term survival of seagrasses (Dennison et al., 1993). Moreover, other parameters such as sediment composition, wave exposure and current velocity need to be considered for successful planting of seagrasses (Koch, 2001; Fonseca et al., 2002). Seagrass restoration can be promoted via seed collection and subsequent dispersal (Orth et al., 1994, 2000) or transplantation of plant material with or without sediment attached to the root system (Fonseca et al., 1998). The long-term success of seagrass restoration projects is still relatively low, with much current effort directed towards understanding the environmental parameters, physiology of various seagrass species, and planting or seeding methods to improve outcomes (see, for example: Kemp et al., 2004; Schenk and Rybicki, 2006; USGS, 2002). Due to the low success rate and ongoing research on the degree of wave attenuation and shoreline protection provided by seagrass beds, seagrass restoration is not yet considered a viable method for shoreline stabilization although restoration technologies may improve in the future.

Vegetated Dunes

In addition to previously described methods, dune creation can provide a system to create or maintain a beach because it adds sand that will nourish the area, with or without structural control. Dunes are established along the backshore region of nourished beach by planting the appropriate species of dune grasses. Sand fencing, in conjunction with dune grass plantings, helps induce baffling and settlement of wind-blown sands (Figure 3-2). Moreover, a dune berm can be created to provide a foundation for dune creation, thus providing a head start in the dune building process.

Harden

Perhaps the most widely applied shoreline technique is to harden the shore or bluff with some type of fixed structure such as a bulkhead, seawall, or revetment (Figure 3-3). The primary goal of hardening the shore is to protect the coast from wave attack by creating a barrier to the erosive forces.

Traditional shoreline hardening design involves methods applied at a local or regional scale, often utilizing local materials such as stone, wood, and concrete, and built using techniques familiar to local marine contractors and property

FIGURE 3-2 A dune beach along Virginia's Chesapeake Bay. Note the fencing and dune grass plantings. The fences and vegetation help to induce baffling and sand settlement. SOURCE: VIMS photo archive. Courtesy of VIMS.

FIGURE 3-3 Shoreline hardening along the coast of Long Island, Long Island Sound. Wood bulkhead (upper left); concrete seawall (upper right); stone revetment (lower left); and gabion seawall (lower right).
SOURCE: Tanski, 2005.

owners. For example, in the northeast, stone walls constructed of local rock have been used with long-term effectiveness, whereas in the mid-Atlantic, wood and concrete bulkheads are used extensively. Since the mid-1970s, stone has become more widely used in the Delaware Bay and Chesapeake Bay region. Wooden walls are common shore structures in the sounds and bays of North Carolina, South Carolina and Georgia. Wooden and concrete walls are common around the sheltered coasts of Florida and Alabama. In Mobile Bay, wood bulkheads have been used so extensively that a "bath tub" effect has been created: Even at low tide there is no beach; the "shore" is a bulkhead (Douglass, 2005a). Along the Mississippi and Texas coasts, both rock and, wooden structures continue to be popular. Wood, concrete, and stone bulkheads are used to harden eroding coasts of the Pacific Northwest.

With continued coastal development—extensive in many areas—the amount of shoreline hardening typically increases, with several environmental effects. Firstly, a properly designed and constructed structure will protect the upland from wave attack and stop shore erosion. Secondly, any provision of sediment from the upland to the beach and nearshore will be blocked, a process sometimes called impoundment (Griggs et al., 1994). In some case, the eroding bank or bluff face

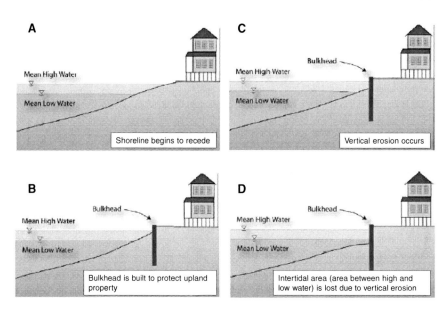

FIGURE 3-4 Progression of a typical response to bay erosion. When the shoreline is receding (A), the homeowner builds a bulkhead to protect the upland property (B) which begins to interfere with the nearshore processes, causing vertical erosion of sediment in front of bulkhead (C), which leads to loss of the intertidal habitat (area between mean high and low water) (D).
SOURCE: Modified from Tait and Griggs, 1990, and Douglass, 2005a,b.

has been stabilized by grading, bioengineering,[4] or both, which also impounds sediment. Thirdly, progressive hardening of an alongshore reach of coast will result in cumulative impacts with regard to loss of sediment. The beach, if present, will begin to decrease in volume and dimension (Kraus and McDougal, 1996; Kraus and Pilkey, 1988). On an eroding shoreline, hard structures such as bulkheads and revetments tend to increase wave reflection and scour, often causing a decrease in the width of the nearshore environment and an increase in water depth (Figure 3-4; Douglass, 2005a,b; Rogers, 2005). These processes can undermine the structure and contribute to erosion on flanking shores, often leading to a pattern of increased erosion—more hardening—increased erosion—and addi-

---

[4]Bioengineering is the use of vegetation, either on its own or in integration with other organic or inorganic structures, to address the problems of erosion. This can be as simple as planting vegetation to help bind and stabilize soils, but also includes the use of more advanced technologies, such as incorporating synthetic geotextiles along with vegetation. Bioengineering is often considered a "soft" armoring method.

FIGURE 3-5 Concrete seawall in residential area adjacent to Lowman Beach Park in Puget Sound, King County, Washington.

tional hardening. As more and more of the shore becomes hardened, the impacts become greater. The cumulative impacts to sheltered coasts include permanent removal of sand from the littoral system creating oversteepened shorefaces, loss of intertidal zones, and intertidal and beach habitat loss.

Bulkheads

Bulkheads are shore anchored, vertical barriers, constructed at the shoreline to block erosion (Figure 3-5). Their popularity, particularly in urban estuaries and sheltered shorelines, has led to broad impacts as adjacent properties are bulkheaded to maintain a consistent shorefront. Douglass and Pickle (1999) have shown that armoring of shorelines in Mobile Bay has resulted in loss of intertidal habitats, such as beach and marsh, as the shoreface becomes progressively armored. This loss may be less rapid or reduced when bulkheads are built landward of the shoreline (Figure 3-6).

Bulkheads built on eroding shorelines affect the shorelines in three ways:

FIGURE 3-6 Landward placement of bulkheads leaving existing marsh and beach intact, at least for the near term.
SOURCE: Tanski, 2005. Courtesy of the New York Sea Grant.

1. Permanent removal of sand from the littoral transport system that nourishes downdrift beaches.

2. Creation of oversteepened shorefaces. In general, sediment deficient shorelines are steeper due to loss of sediment, but bulkheads accelerate this process. When nonbreaking waves impact a bulkhead, the bulkhead reflects close to 100 percent of the wave energy. The wave essentially doubles in height at the structure and as it recedes, considerable forces are exerted on the toe of the structure, creating scour which over-steepens the beach.

3. Reduction or elimination of the intertidal shore as the shoreline erodes with a resultant loss of habitat and recreational access. An eroding shoreline normally maintains a certain profile shape as it migrates landward. If a structure is placed landward of that eroding shoreline, the water will eventually migrate to the toe of the structure and fronting marsh or beach will be lost. This is diagrammatically depicted in Figure 3-4.

Bulkheads may be constructed of wood, concrete, vinyl, or steel, and can be freestanding or have a series of tiebacks for stability. When properly designed and constructed, bulkheads can greatly reduce or temporarily eliminating shoreline retreat at a site. Scour from the reflected waves will increase the depth of water at the bulkhead base. Therefore, stone or other riprap is often placed at the toe to absorb some of the wave energy. If the bulkhead is constructed at the shoreline, the area landward of the bulkhead is typically filled, and the marsh or beach is converted to uplands.

Seawalls

Seawalls (Figure 3-7) differ from bulkheads in that they are designed to withstand greater wave energy and are more likely to be constructed on open coasts to protect against ocean wave climates. They are most often constructed with cast-in-place concrete; other materials such as timber are rarely used. These structures can be vertical, curved or stepped to help divert or redirect wave energy. A sloped face may reduce the effect of toe scour but conversion of habitat will still occur if erosive forces continue to remove sand.

Revetments

Revetments armor the slope face of the shoreline (Figure 3-8). They are commonly constructed with one or more layers of graded riprap but can also be constructed with precast concrete mats, timber, gabions (stone-filled, wire-mesh baskets), and other materials. Although revetments will successfully stop erosion when designed and constructed properly, many projects are haphazardly constructed using available materials (e.g., broken asphalt, car bodies, concrete, building rubble, and other waste materials) with little planning. Such structures

FIGURE 3-7 Concrete seawall with groins on James River in Newport News, Virginia. SOURCE: Hardaway and Byrne, 1999. Courtesy of VIMS.

often perform poorly and, in some cases, can accelerate erosion by destabilizing banks. As with bulkheads and seawalls, revetments are frequently placed at the shoreline and backfilled to create a limited sloping backshore. If a revetment is placed on an eroding shore and erosion continues to move the shoreline towards the revetment, the original intertidal habitat and beach will be lost to open water.

Breakwaters

Breakwaters consist of a single structure or a series of units placed offshore of the project site to reduce wave action on the shoreline. The structures are composed of various types of materials but usually employ what is "locally" available. Rock is typically used for construction and has been shown to be very durable when properly designed and installed. Other materials have been used with varying degrees of success, including broken concrete, formed concrete, and tires. A breakwater or breakwater system may or may not include the addition of sand to the system depending on the design, site conditions (whether there is abundant sand in the reach), and the level of shore protection required (a more detailed discussion is provided later in this chapter).

FIGURE 3-8 Stone revetment shortly after construction on the Potomac River, Virginia, and a cross-section of elements necessary for proper stone revetment design.
SOURCE: Hardaway and Byrne, 1999. Courtesy of VIMS.

Sills

Sills (Figure 3-9) are generally semicontinuous structures built to reduce wave action and thereby preserve, enhance, or create a marsh grass fringe for shore erosion control. The sill is often built along an existing marsh fringe to maintain its integrity and enhance the protection afforded by the marsh in controlling erosion on the adjacent upland. The addition of sand with marsh grass plantings provides a stable marsh fringe system in low to moderate wave energy environments. Breaks or windows in the sills are recommended to allow the ingress and egress of marine fauna.

Building a sill system requires encroachment bayward or riverward, usually beyond Normal High Water or Mean High Water (MHW), constituting the property limit in most states and complicating the process for obtaining permits for installation. There is often a trade-off of habitats in constructing a sill system. The eroding bank, narrow beach and nearshore are converted to a stable bank, marsh and stone sill. In addition, the sill system may reduce the sediment supply to adjacent shores.

The sill structure is often composed of stone, but treated wood and other materials are also used. The sill is usually low, designed to trip or break storm waves before they cascade across the marsh fringe, thus dissipating wave energy before it reaches the upland bank and minimizing marsh toe erosion. A typical sill is illustrated in Figure 3-9, where sand fill has been added to create a gentle seaward slope from the bank to the back of the sill structure at mid-tide elevation. Once the sill has been properly constructed, appropriate wetlands grasses are planted to establish or supplement the marsh.

Trap or Add Sand

For landowners in particular, creating and maintaining a beach and dune for shore protection is often the most desirable option. Trapping and adding sand or gravel creates an effective shore planform and cross-section for shore protection. Structures installed perpendicular to the shore (e.g., groins) and parallel to the shore (e.g., breakwaters) are used to trap sand, frequently in conjunction with projects that add sand to the shoreline. Groins will reduce the volume of sand transported downstream, potentially depriving these areas of sand needed to maintain a beach.

Beach Nourishment

Beach nourishment is the addition of sand to a shoreline to enhance or create a beach area, offering both shore protection and recreational opportunities. Beach nourishment studies, design, and projects have been focused mostly on ocean coasts (NRC, 1995a). On sheltered coasts, beach nourishment is most often used on public lands to promote recreational use. Sometimes, beach nourishment

FIGURE 3-9 Stone sill examples from St. Mary's River, St. Mary's County, Maryland, and North Carolina Patches of existing marsh were incorporated into the plan.
SOURCE: VIMS shoreline photo archive, and North Carolina Coastal Federation. Courtesy of VIMS and the North Carolina Coastal Federation.

projects are combined with groins or breakwaters systems to help retain the added sand.

Although addition of sand to a beach can effectively create a protective beach, a periodic maintenance cycle is usually required with this technique, which may prove unwieldy for the individual property owner. The source and quality of beach material is a significant part of the overall cost of a beach nourishment project. These sources may include dredged sand from maintenance of nearby navigation channels or sand mined from upland borrow pits. The latter option is typically more expensive, but may be the only sand source available in some shoreline situations. Sand quality is recommended to be at least as coarse as the native sand (NRC, 1995a).

Beach nourishment replenishes sand lost to erosion and protects the adjacent upland from storm wave impacts. The beach can be "engineered" so that its dimensions, both cross-section and shore planform, will provide a level of protection for known or predicted storm events (NRC, 1995a). Sand added to hold the bluff in place and protect the dunes is often referred to as the design beach. Additional sand, the sacrificial beach, provides temporary protection that is expected to be lost during major storms. The addition of vegetation in the form of dune grass plantings provides an added level of protection and an important habitat component (Figure 3-10).

The addition of sand to a coastal shoreline, at least initially, forms a covering on the existing shore zone and nearshore region. At first, the nearshore benthic community is replaced by an intertidal and supratidal beach and dune. Some states are reluctant to allow this option because of the effect on native habitat, but other states (for example, Virginia code section 10.1-704) encourage the addition of clean sand to the littoral system because existing natural beaches along many sheltered coasts are rare, typically low and narrow, and often too small for effective shore protection.

Groins

A groin is a barrier-type structure, used on a variety of coasts including sheltered shores and open coasts, that traps sand by interrupting longshore sand transport. Groins extend from the backshore into the littoral zone and are normally constructed perpendicular to the shore out of concrete, timbers, steel, or rock (Figure 3-11a). A structure is classified as high, medium, or low energy depending on the percentage of littoral drift that the groin interrupts.

By capturing sand, the groin reduces the sediment supply to the downdrift beach, potentially triggering erosion or accelerating the rate of erosion on the downdrift side of the groin. This accelerated erosion will result in narrowing and loss of beach habitat. To compensate, additional sand can be added to the groin project. Figure 3-11b shows one such project where sand was imported to the site and a beach was constructed to replace a shoreline of graded riprap and broken

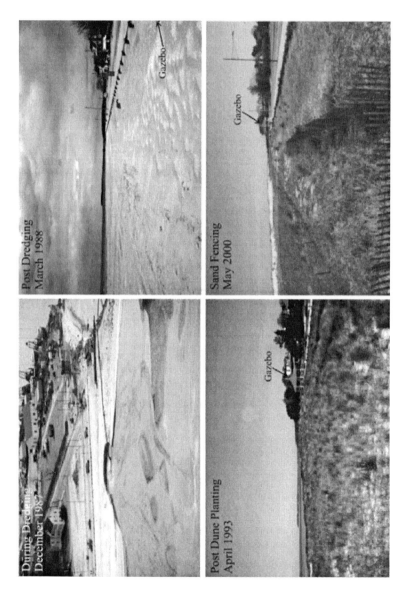

FIGURE 3-10 Cape Charles, Northampton County, Virginia, beach fill and dune field development. SOURCE: VIMS shoreline photo archive. Courtesy of VIMS.

FIGURE 3-11 Groin field on the Rappahannock River, Virginia, with an adequate sand supply to provide a protective beach zone to upland property (A) and an inadequate supply along the shore reach (B).
SOURCE: Hardaway and Byrne, 1999. Courtesy of VIMS.

concrete. From an aesthetic perspective, groins create a more natural shore face than bulkheads or revetments, but may present a hazard for boating and other recreational pursuits.

Breakwaters

Breakwaters may be designed as either detached or attached to the shoreline (Figures 3-12 and 3-13). Detached breakwaters sit offshore at a distance that precludes a connection to the beach. A salient (seaward growth of the shoreline) will form in the lee of the structures, but often requires beach fill. An attached or headland breakwater, as the name suggests, is connected to the mostly sandy beach shoreline, often with beach fill. The headland breakwater system is composed of a series of pocket beach and breakwater units designed to maintain the adjacent beach in a predictable shore planform for shore protection or to create a recreational beach. Recreational beaches are often larger than what might be needed for a shore protection application. Both the desired width and length of a recreational beach may exceed what is necessary for shore protection.

If there is sufficient sand in the littoral system, beach fill requirements may be minimal if any. On sheltered coasts, a scarcity of sand often leads to bank erosion. Therefore, a breakwater system will usually include a beach-fill component. As with groins, the downdrift coast will sometimes become deprived of sediment. To avoid excessive impacts, these conditions should be evaluated and addressed during the design process. Some options are shown in Figure 3-14 for breakwaters exposed to different annual storm wave climates. This type of shore management is well suited to relatively long segments of coast (long reach) with defined geomorphic boundaries.

Composite Systems

Composite systems incorporate elements of two or more methods discussed above in order to provide long-term shore protection to a coastal reach. The purpose is to utilize complementary elements of techniques to best fit the specific situation. An example of a composite system is shown in Figure 3-13 and includes a headland breakwater system with stone breakwater units, sand fill, and vegetative stabilization of the backshore and graded bank. The system transitions from a low reef breakwater, shown at the top of the photo, to the stone revetment shown at the bottom. This particular system performed exceptionally well under direct storm surge and wave attack from Hurricane Isabel, preserving the heavily vegetated interface between the backshore and base of the bank.

Another composite system is shown in Figure 3-9, showing typical application of a sill in Chesapeake Bay. A stone sill, clean sand fill, and wetlands vegetation combine to provide a stable marsh fringe. In other cases, sand with wetlands plantings are used in combination with small stone groins (Figure 3-1, middle

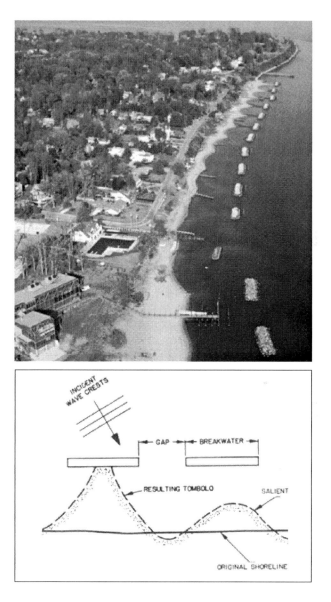

FIGURE 3-12 A detached breakwater system at Bay Ridge on Chesapeake Bay in November 2002. Definition sketch of beach morphology of tombolo and salient as they relate to offshore breakwaters (after Dally and Pope, 1986).
SOURCE: VIMS, 2006. Courtesy of VIMS.

64

FIGURE 3-13 Kingsmill on the James, James City County, Virginia. Shore protection system utilizing primarily headland breakwaters and beach fill with wetland vegetation, bank grading with upland vegetation, and an interfacing low-crested breakwater and revetment. SOURCE: Hardaway and Byrne, 1999. Courtesy of VIMS.

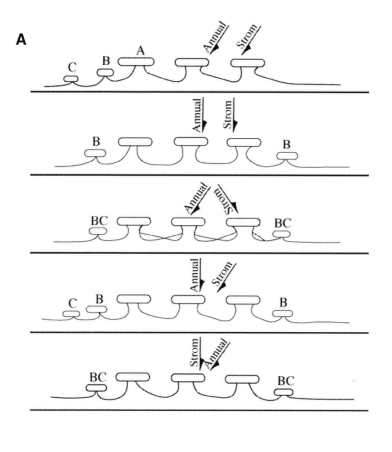

FIGURE 3-14 Various wave climates (A) found on straight stretches of sheltered shore-lines and the corresponding cross-sections (B) of the engineered breakwaters.
SOURCE: Hardaway et al., 1993. Courtesy of the American Society of Civil Engineers.

image). The use of rock, sand, and plants in various combinations has been shown to be an effective approach to shore erosion over the long term.

Beach nourishment projects have frequently included structures to help retain the sand. For example, breakwaters can maintain a beach in an equilibrium planform that is effective in dissipating surf-zone energy during storm events. A series of breakwaters and beach fill can be used to address a long segment of shoreline as a breakwater system, additional resilience to storm events can be gained by vegetating the backshore region of the breakwater system, particularly if it lies against an upland bank that has been graded (Hardaway et al., 2005).

These "softer" approaches, such as the vegetated shoreline or nourished beach, often require maintenance over time including replacing sand and plantings. Also, the base of bank might need to be regraded if the system experiences a storm that exceeds the design specifications.

Headland Control

When erosion is managed on a reach basis, headland control can be used to reduce erosion rates on long stretches of a sheltered coastline. Usually breakwaters are used in conjunction to reduce the linear feet of structure needed and keep the overall costs down. Headland control involves either the reinforcement or accentuation of an existing geomorphic feature or creation of an artificial headland that allows the adjacent, relatively wide, embayment to achieve a stable configuration (Figure 3-15).

Nontraditional and Innovative Methods

Numerous shoreline erosion control devices have been developed over the years with varying degrees of success; an extensive listing of devices is maintained by Duke University (Duke University, 2005). For example, geotextiles are being developed that form a water permeable barrier that holds back sediment. The geotextile may be used to create a tube filled with dredged material or sand—it has the advantage of being flexible and can be arranged to provide a more optimal configuration (Eurosion, 2004). This relatively new technology will require further testing to determine efficacy and longevity in a range of environments. The U.S. Army Corps of Engineers, under Section 227, funds projects which provide for the demonstration and monitoring of innovative techniques at site locations around the country (USACE, 2006a). Techniques such as breakwater and sill systems in the United States were considered innovative and new in the 1980s, but their efficacy has been documented and such systems are widely used in some regions. Therefore, developing new and innovative methods and techniques is important to improve and maintain sheltered coastal environments. One promising approach under development is the concept of sand and shore management.

FIGURE 3-15 Placing widely spaced breakwaters and allowing adjacent embankments to erode and evolve into equilibrium embayments can be a cost-effective method of reach management, as seen at (A) Hog Island, James River, Virginia, and (B) Westmoreland County, Virginia, in Potomac River, installed in 1998.
SOURCE: Hardaway and Gunn, 1999. Courtesy of the American Society of Civil Engineers.

DESIGN ELEMENTS AND CRITERIA

The vast majority of options for addressing erosion of sheltered coasts are designed to provide a level of protection that balances the desire to halt erosion with the costs, both financial and environmental, of the protective strategy. This section provides a general outline that discusses design elements including the level of protection, and damage and risk. The fundamental causes of shore erosion and how these causes affect how erosion is addressed are discussed in the Chapter 2 section on "Implications of Geomorphic Setting for Erosion Mitigation Strategies."

Design methods can be found in numerous publications such as "Shore Protection Manual: Low Cost Shore Protection, a Guide for Engineers and Contractors," (USACE, 1981). Many of these techniques primarily target open-coast projects although some address sheltered coasts. Assessment of sheltered coast protection has some specific elements, but many of the general principles apply to a lesser or greater degree, depending on the site. When designing erosion mitigation structures, two key elements must be taken into consideration Firstly, the problem must be adequately defined and secondly the desired outcomes must be specified (USACE, 2000: The Coastal Engineering Manual).

Regardless of the size of the project, some level of design needs to be employed. For example, in the case of the waterfront lot owner and local contractor, the design may be strictly empirical, pointing to "successful" structures nearby: An extensive wave climate analysis is not necessarily warranted. At high fetch exposures, the potential impacts from impinging wind-driven waves can be a critical factor in establishing a marsh fringe, maintaining a beach, or correctly determining the size of stone for shoreline methods. The greater the fetch, the larger the potential wind-driven waves (i.e., storms), the greater the required level of protection and the greater potential impacts to adjacent lands by a shore protection method.

Level of Protection

The level of protection recommended in a given area tends to be subjective. Broadly, the context of level of protection is "what got us through the last storm." Local preferences are typically the methods adopted by local marine or shoreline contractors to abate shore erosion. Features such as wave cut upland scarps, wrack lines and water marks resulting from hurricanes and extratropical storms are readily identified by local contractors who use them to offer a level of protection sufficient to weather a "Storm of 1991" or the last major hurricane. Therefore the level of protection may be defined as the horizontal and vertical dimensions required for a shoreline project to protect the coast from erosion during the design storm.

Quantifying storm waves and storm surge impacts and their return intervals or frequency is an issue that needs to be addressed when designing erosion mitigation procedures. Most coastal localities have Federal Emergency Management Agency (FEMA) Flood Insurance Rate Maps (FIRM) with 100-yr storm surge levels (Coulten et al., 2005). This means that there is a one percent chance that the stated water level will occur in any given year. The added component of storm waves is also shown as V zones on the flood maps. Storm waves on top of the storm surge increase the height of the water that impacts the coast. This information is generally accessible from the locality and should be referenced by the waterfront property owner. More detail, such as the 50-yr and 25-yr storm surge levels, is provided in locality specific FEMA studies and reports.

In low lying areas that will be readily flooded by the 100-yr event or even a lesser storm, the question of level of protection needs to be evaluated (see Figure 2-8). A shore protection method can be installed against the high bank to address the 100-yr event with a relative degree of straightforwardness whereas a low bank requires some potentially difficult decisions. It may be impractical to bring a stone revetment up to the 100-yr level since it might be several feet higher than the adjacent bank. Aesthetics might also be a consideration. For a breakwater system with beach nourishment, creating a dune at the design level provides additional protection and if overtopped, more sand can be added. All shore protection systems may require maintenance. The key is to have the system remain intact even if overtopped during a storm.

As the period of record lengthens, quantification of risk becomes increasingly statistically significant. The quantification can generally be portrayed by design professionals, engineers, and competent marine contractors through the design process. However, at the waterfront lot level, this information is often not supplied by the contractor. The contractor can point to "long-standing structures" in the vicinity to show success with the level of protection implied. Although this appears to demonstrate shore protection effectiveness, it lacks a quantitative basis for evaluating performance.

There is always the possibility that the level of protection will be exceeded by an event greater than the "design storm" for which the mitigation was designed. The practicality of expanding levels of protection, for instance to withstand the 500-year storm event, is not easily determined, especially considering the additional costs associated with increased levels of protection.

The need for shore protection and the required level of protection is often driven by the threat to infrastructure. In addition to the potential loss of land, erosion threatening loss of a house or a road frequently provides the motivation for mitigation (Figure 3-16). Development at or near an unstable slope or eroding shore will force shore protection action in the future. Therefore, consideration should be given at the local planning level to the consequences of development in highly eroding areas. For example, developers may be required to include shoreline management plans to obtain construction permits.

Damage and Risk Assessment

The level of protection employed will translate to the amount of risk or damage the property owner is willing to accept or incur and the amount budgeted for installating protection. Some level of damage may be deemed acceptable depending on the size of the project and the value of the property to be protected.

Although generally used interchangeably, there is a difference between shore erosion control and shore protection. Shore erosion control does not provide a specific level of protection. In other words, doing just about anything will provide some erosion control over the current condition. Many unproven devices will provide some shore erosion control but may not provide shore protection.

Shore protection is defined by the level of protection provided based on an analysis of site conditions (i.e., the design). The protective structure is typically designed to withstand a given intensity of a particular event, such as a storm. This is referred to as the "design event." If a more intense event occurs, the level of protection will be exceeded. Overtopping a revetment by surge and wave may create a wave-cut scarp across the adjacent bank or bluff (Figure 3-17). If

FIGURE 3-16 The result of bluff erosion past the point of being critical.
SOURCE: VIMS shoreline photo archive. Courtesy of VIMS.

the revetment stays intact there may not be a problem as long as the bank face remains stable. If the structure itself fails, particularly early during the storm event, then the bank will fail and infrastructure may be threatened or damaged (Figure 3-18).

Risk can be related to return frequency of the design condition which may be measured in terms of a "design wave," the level of water (storm surge) or waves anticipated during a specified time interval. For example, if a property owner wants to protect the shoreline from high water for 10 years, the designer might choose a 10-year design wave condition (see Table 3-1; Figure 3-19). The chance of experiencing the design wave during the structure's first 10 years would be 65 percent. If the structure lasted for 25 years, there would be a 34 percent chance of the design wave occurring during the structure's lifetime. In most cases, the project life is designed for 25 years (25-year design condition), with a 64 percent chance of failure during that time interval.

EROSION CONTROL STRATEGIES IN APPLICATION

The following hypothetical example is offered to illustrate how site conditions affect the range of effective erosion control options available to the homeowner.

Post Isabel
October 2003

FIGURE 3-17 Overtopping of stone revetment on south side of James River in Isle of Wight County, Virginia. Top of structure is at 2 meters (approx. +8 feet) MLW. Note significant bank scarping due to Hurricane Isabel when the combination of storm surge and wave runup reached 4 meters (approx. +12 feet) MLW. The structure and upper bank face are still intact.
SOURCE: Hardaway et al., 2005. Courtesy of VIMS.

Homeowner's Dilemma

A homeowner lives on a 200-ft waterfront parcel on a tidal creek. The shore-front consists of a narrow beach backed by a 10-ft high eroding bank. No buildings are immediately threatened but with every northeaster the homeowner loses about a foot of land. The neighbor on the right (Neighbor R), south and downstream of the homeowner, has a wood bulkhead and graded bank. The neighbor on the north side (Neighbor L) has a similar shorefront as the homeowner, but with less erosion on the upstream side where the bank is fronted by a marsh fringe (see Figure 3-20). The homeowner decides to investigate options for stemming the erosion of her property and hires a consultant. The consultant analyzes the site conditions and offers the 4 options described below.

Consultant's Site Analysis

The shoreline is on a slight headland that is exposed to the northeast. To the north and southeast, the fetch is less than a mile, but the northeast opens to a sound, about 3.0 miles across to the facing shore.

FIGURE 3-18 (A) Stone revetment built with only one layer of undersized armor stone on too steep a slope and (B) its failure after a modest storm event.
SOURCE: Hardaway and Byrne, 1999. Courtesy of VIMS.

There is a 2-ft tidal range and storm surges of 2 ft or more above normal can be expected every 3 to 5 years. Larger, less frequent storms will cause storm surges of over 5 feet. Bank erosion is associated with storms and associated high water and wave action. Neighbor L has a similar problem that becomes less severe toward the north because the property is more sheltered from the northeast exposure. The bank face on the opposite side of the lot is fairly stable with heavy vegetation but a significant wave-cut scarp along the base. The marsh fringe on

TABLE 3-1 Project Life vs. Risk

Project Life (years)	Design Condition Return Percent	
	10 yrs	25 yrs
1	10	4
2	19	8
5	41	18
10	65	34
15	79	46
20	88	56
25	93	64
30	96	71
40	99	80
50	99	87

NOTE: Data from this table are derived from the graph in Figure 3-19.
SOURCE: Data from British Standards Institution (1991).

the far side of Neighbor L's lot offers some wave buffering during storms. However, boat-wake impacts may increase as development continues along the creek, causing more erosive damage to the marsh fringe.

Options for Addressing Erosion

Option #1—take no action. Since the house is 75 ft from the bank, infrastructure will not be at risk for many years given an average erosion rate of less than one ft per year. Loss of property and landscaping (notably trees) will continue. This will have no direct costs for the homeowner.

Option #2a—create marsh fringe vegetation. The marsh fringe on Neighbor L's lot indicates that it may be possible to build a marsh that extends further downstream in front of the homeowner's lot. There are many places along the upper reaches of the creek where just trimming trees and planting the existing substrate could significantly enhance a protective marsh fringe. However, the 3 mile fetch to the northeast at the homeowner's site will make the marsh vulnerable to storm-driven waves, offering minimal erosion protection during major storms. In addition, the marsh fringe will require ongoing maintenance. Cost to the homeowner is low, but there will be ongoing maintenance costs.

Option #2b—create marsh fringe vegetation with sill. To protect the new marsh from wave exposure a sill will be installed to attenuate wave action. The sill, typically composed of rocks, will run parallel to the shore at a distance to match the desired width of the created marsh. Sandy fill will be placed behind

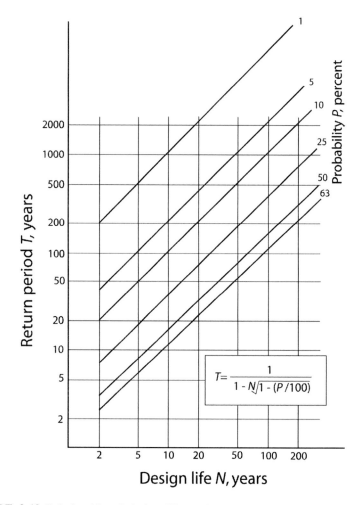

FIGURE 3-19 Relationship of design life to the return period and probability of exceedence. T is the return period of a particular extreme wave condition in years. P is the probability of a particular extreme wave condition occurring during design life N years. SOURCE: Derived from British Standards Institution, 1991.

the sill to raise the backshore to establish a wetlands planting terrace and provide greater storm protection. The bank could be graded to adjoin Neighbor R's bulkhead.

The sand fill level will be about 3 ft above MHW at the base of the graded bank. The sand fill will be graded to form a 10:1 slope to intersect the back of the sill at about mean tide level (MTL). This is about the lower limit of tidal marsh

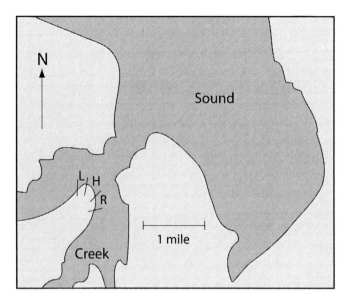

FIGURE 3-20 Map illustrating the exposures of the hypothetical homeowner's property (H) and orientation with respect to neighboring properties on the left (L) and right (R).

growth in this region. The sill will be about 40 ft creekward of the base of the graded bank. The bank may need to be graded to an 8:1 to blend with neighbor R's bulkhead and reduce encroachment onto state-owned creek bottom. The sill system will continue in front of neighbor L's lot but the bank grading would not be needed at the side that contains a marsh fringe. Some of the original marsh will be affected, but the new marsh fringe should compensate for the loss. Some of the sandy graded bank material will be used for the wetlands terrace but any excess will be hauled offsite. There may be some maintenance required after storm events, especially in the first few years until the marsh becomes established. Native shrubs will be planted on the bank face to create a riparian buffer and minimize the need to fertilize (compared to grass), reducing the nutrient input into the creek and state waters. This option is very expensive for the homeowner who must cover costs of construction, maintenance, and obtaining a permit to place a sill on state-owned creek bottom. Obtaining a permit may be difficult because of multiple levels or regulatory review. Also, this option will require the cooperation of neighbor L, but it preserves the visual landscape which is a high priority for both neighbor L and the homeowner.

Option #3—extend the bulkhead. Installation of a bulkhead will mostly likely result in the loss of the narrow beach at the base of the homeowner's bank.

The life expectancy of the bulkhead should be at least 20 years, possibly more depending on the durability of the bulkhead material. Neighbor R's bulkhead is about 3 feet above mean high water (MHW) and is overtopped during large storms. There is evidence of bank scarping by wave action, but the integrity of the bank slope is intact and is repaired with some minor fill and vegetation. If the homeowner and Neighbor L were to install bulkheads at the same time with the same contractor, the cost per property would be lower because they could split the contractor's mobilization and demobilization costs. Cost to the homeowner is moderate, no federal permit will be required, and this option is preferable to neighbor R. The project design life of the bulkhead should be at least 20 years, possibly more depending on the durability of the bulkhead material.

Option #4—install a stone revetment. As with a bulkhead, installation of the revetment will mostly likely result in the loss of the narrow beach at the base of the homeowner's bank. The revetment should be at least as tall as neighbor R's bulkhead. The bank would be graded by cutting the top back and pushing it creekward to create a subgrade. The minimum required bank slope is 2:1, but lesser slopes, say 3:1, would provide more effective wave attenuation during large storms and reduce bank scarping. If neighbor L's lot was included in the project, the adjacent bank would be graded but only enough to meet the stable vegetated bank at the other end of L's property. Here the revetment would continue along a new subgrade in front of the stable bank face. Some of the existing marsh would be covered by the structure, potentially requiring some form of compensation for loss of wetland. Cost to the homeowner is moderate to high. Although construction will require a permit, the permitting should be straightforward and unlikely to cause a major delay in the project. The project design life of the rock is 50 years or more and the integrity of the structure depends on quality construction.

In addition to showing how the site conditions affect the suitability of erosion control measures, this hypothetical case indicates some of the choices that face a homeowner with regard to cost, permitting, and potential changes to the landscape. The decision-making context for addressing erosion is further explored in Chapter 5.

FINDINGS

- Strategies that address erosion, other than land use controls, can have cumulative impacts to sheltered coasts. These include permanent removal of sand from the littoral system, creating oversteepened shore faces, loss of intertidal zones, and habitat loss.
- Managing land use has long-term individual and cumulative benefits that extend beyond those produced by other types of erosion control.
- There are different strategies for shore protection, but the final design choice depends on landowner's goals, level of protection, risk, site assessment,

and expense. These elements may have differing priorities for a given project but all are relevant to achieving optimum costs and benefits. Matching any of the many approaches to the appropriate setting then becomes the fundamental challenge.

• Many engineers, contractors, and property owners are unaware of the range of options available for controlling erosion.

4

Mitigating Eroding Sheltered Shorelines:
A Trade-Off in Ecosystem Services

C oastal engineering projects designed to protect the shoreline from erosion
     focus mainly on the need to preserve assets such as buildings, roads, rail
     lines, and lighthouses. The impacts of these coastal engineering projects
on organisms and natural processes usually receive less attention, if any. While
coastal structures may protect shorelines for a limited time, they also alter the
coastal environment. Surprisingly, little attention has been paid to the ecological
consequences of installing structures to mitigate shoreline erosion (Airoldi et al.,
2005). This chapter focuses on the trade-off of ecosystem services associated
with shoreline protection methods, i.e., the loss of ecosystem services of natural
coastal communities along sheltered coasts that are being protected from further
erosion and the gain of ecosystem services associated with man-made structures
built to protect the shoreline from erosion.

Coastal ecosystems provide a variety of marketable goods (e.g., fishes,
fibers, seaweeds, crabs, sand) as well as processes (e.g., climate regulation, wave
attenuation, removal of nutrients, contaminant sequestration, maintenance of biodiversity) that allow humans to thrive (NRC, 2005). These goods and ecosystem
processes that benefit humankind are often referred to as ecosystem services.
For example, the quality of life in coastal towns and villages is enhanced by ecosystem services such as nutrient uptake, habitat and food production provided by
coastal plant communities. These services support commercial fisheries, as well
as the recreational use of the shorelines and adjacent waters. Monetary values
have been assigned to individual ecosystem services; (Costanza et al., 1997; but
see Nature, 1998, for a comment on Costanza et al.), however, because of the
challenges posed by the valuation process of ecosystem services (NRC, 2005),

this chapter brings awareness to ecosystem services lost and gained by the mitigation of shoreline erosion along sheltered coasts without attempting to incorporate specific monetary values.

Natural shorelines are normally dynamic and undergo natural cycles. Vegetated shorelines experience a seasonal fluctuation in plant biomass and sandy beaches are constantly reworked by waves. These cycles allow a system to maintain equilibrium around some degree of disturbance followed by a period of recovery. Each time a natural system is disturbed, some of its ecosystem services are lost. If the level of disturbance is low in magnitude and frequency, the ecosystem and its services will recover over time. Recovery time is longer when the magnitude of the disturbance is higher. For example, a wind event may cause sediments to be resuspended in shallow waters leading to high turbidity levels no longer allowing oysters to filter. The ecosystem service of filtering water is therefore reduced or lost until turbidity levels return to acceptable levels (usually in the matter of hours). In contrast, if a hurricane buries an oyster reef, its ecosystem service of filtering water is lost until the reef is reestablished (years or even decades later).

The current trend of enhanced shoreline erosion and subsequent shoreline protection (see Tables 4-1 and 4-2) leads to disturbance along sheltered shorelines

TABLE 4-1 Extent of Shoreline Armored in California

Year	Length	Percent of Coast	Source	Other Information
1971	42.4 km (approx. 26.4 miles)	2.4	USACE	armored exclusive of breakwaters and groins
1978	100 km (approx. 62 miles)	5.7	Habel and Armstrong, 1978	"protected by engineered structures"
1985	168 km (approx. 104 miles)	9.5	Griggs and Savoy, 1985	seawalls, revetments, breakwaters
1992	208 km (approx. 129 miles)	11.8	Griggs et al., 1992	hard, engineering structures

SOURCE: Griggs and Patsch (2004).

TABLE 4-2 Armoring (By County) in Heavily Populated Central and Southern California

Location	Length of Shoreline	Percent Armored
northern Monterey Bay	14.4 km (approx. 8.95 miles)	77
between Carpenteria and Ventura	28.8 km (approx. 5.97 miles)	77
from Oceanside to Carlsbad	12.8 km (approx. 7.95 miles)	86
from Dana Point to San Clemente	12.8 km (approx. 7.95 miles)	~100

SOURCE: Griggs et al. (1992), reported in Griggs and Patsch (2004).

no longer allowing the natural system to become reestablished to its original state. Instead, a new equilibrium is reached altering the ecosystem services provided. In some highly populated areas, 75 to 100 percent of the shoreline is armored (Table 4-2) and the long-term ecological consequences are unknown.

Practices commonly implemented to mitigate shoreline erosion (e.g., beach nourishment; construction of seawalls, breakwaters, and groins) not only introduce a certain level of disturbance changing the equilibrium of the coastal ecosystem but usually also introduce new substrate such as concrete, rock, wood, or coarser sand. As a result, organisms that can benefit from this new substrate may appear in the ecosystem. Therefore, as an ecosystem service of the original (eroding) system is lost (e.g., habitat provided by fallen trees), an ecosystem service of the plant and animal community taking advantage of the new substrate may be gained. We refer to this as a "trade-off" in ecosystem services. The alteration of a small section of the shoreline via the construction of a single structure to control erosion at a specific site alters the ecosystem services in the immediate area. The construction of many structures in relative close proximity to each other within a body of water may lead to the shift of ecosystem services of the entire ecosystem. Therefore, the cumulative effect of shore stabilizing structures on ecosystem services is also considered in this chapter.

ECOSYSTEM SERVICES PROVIDED BY
NATURAL COASTAL SYSTEMS

This consideration of coastal systems focuses on those that are subject to erosion on sheltered shorelines. These are either constructional coastal features such as beaches, dunes, mudflats, and vegetated communities (both intertidal and subtidal), and erosional landforms such as bluffs, which contribute sediment to sheltered coasts. While some sheltered shorelines include hard rock outcrops, the erosion of features such as rock cliffs or shore platforms on sheltered coasts is considered a slow process and one unlikely to result in the need for the protective shoreline measures that are the focus of this study.

Beaches and Dunes

Ecosystem Services

For dunes and beaches (see Figure 4-1 and Chapter 1 for description), the ecosystem services provided depend on the structure and local environmental factors such as climate, salinity, turbidity, and wave energy to which they are exposed. Ecosystem services commonly listed for dunes and beaches include:

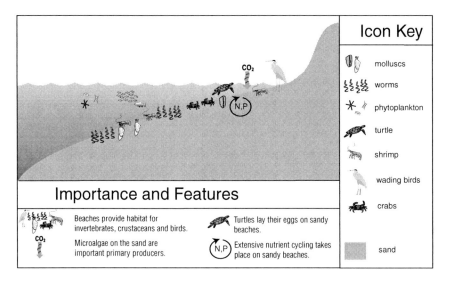

FIGURE 4-1 Conceptual diagram of a beach emphasizing its importance along sheltered coasts and outlining processes that occur on beaches.

Habitat

These sheltered coastal environments provide habitat for a variety of organisms. The shallow refuge areas of shoreface environments provide suitable conditions for pupping by some shark species. Some turtle species nest on upper beaches and within low dune areas in bays and estuaries—an important consideration if erosion mitigation measures are proposed because many populations are endangered or threatened. The subaerial environments can also provide important habitat for mammals, including rabbits and small rodents (Van Aarde et al., 1996), as well as for nesting birds (Lafferty, 2001).

Nutrient Uptake

Vegetation on the upper beach and in dunes is frequently nutrient limited. Many dune soils are deficient in macronutrients though levels of calcium may be high. Experimental work shows that addition of various combinations of nutrients to dune soils results in growth of different combinations of dune plants, especially grasses (Packham and Willis, 1997). In general, nutrient deficiency promotes a diverse plant community. Grazing by herbivores (especially where focused on some of the nutrient-fixing dune plants) can also impact nutrient availability. Sandy substrates such as those found in beaches are quite effective in nutrient cycling (Rasheed et al., 2003; Ehrenhauss et al., 2004); therefore, any alteration to the substrate such as accumulation of fine particles landward of a breakwater will affect this process.

Food Production

Beaches support an extensive trophic structure, mainly in the form of infauna (animals that live in the sediment), from bacteria and microalgae to molluscs, crustaceans, and shorebirds.

Wave Attenuation

The role of beaches and dunes in protecting interior areas from wave attack is very dependent on magnitude of wave attack. Beach form can adjust to dissipate wave energy provided there is sufficient sediment supply. Such adjustment has been well documented in response to seasonal changes in wave conditions (Wright et al., 1979).

On sheltered shores, boat-wake wave energy may also be seasonal and interaction between these and the wind-wave regime, especially in areas of limited sediment supply, can limit the ability of the beach to adjust.

Sediment Stabilization

For beaches and dunes, the provision of this ecosystem service is very dependent on presence of vegetation (see below). Nutrient limitation can result in low vegetative cover in dunes and active erosion of beaches along sheltered shorelines indicates that this service is impaired in many areas.

Recreation

This is one of the most socially important services provided by beaches and dunes. Many sheltered shorelines, especially in estuaries, are close to urban areas, and large numbers of people take advantage of beaches and public access opportunities. Pressure from offroad vehicles as well as continual trampling by foot traffic can lead to blow outs (amphitheater-shaped arenas where the wind has carved away the dunes) on dunes (Hesp, 2002).

Raw Materials

Excavation of sand and gravel from beaches and dunes for aggregate has been a problem on some open coasts (Borges et al., 2002) but is likely not an issue for sheltered shores where the volume of material available is lower and inefficiently renewed due to lower rates of sediment transport.

The Impact of Shoreline Stabilizing Structures

Erosion of beaches and dunes is a natural process and part of the dynamics of coastal areas. Eroding beaches and dunes still provide many of the ecosystem services described above. Fallen trees along the shoreline also serve as additional habitat for terrestrial and aquatic organisms due to the structural complexity they bring to the environment. The provision of sand by eroding dunes also benefits species that require sandy substrates (see case study mentioned later in this chapter). In most instances shoreline structures are placed in order to address local erosion and protect private property, in some cases preventing the loss of an ecosystem service (e.g., beach access to nesting turtles). However, the placement of erosion control measures frequently have unintended consequences in reducing the ecosystem services provided both by proximal and distant beaches and dunes.

Beaches

Some shore stabilizing structures will prevent further erosion of sandy shorelines, maintaining the beach habitat and other services. However, when structures interrupt longshore sediment movements, they may result in loss of beach ecosystem services in downdrift locations, such as turtle nesting, recreational opportunities, and wave energy dissipation. This effect is well illustrated in a number of shoreline studies (e.g., Pilkey and Wright, 1988).

Dunes

Sheltered coast dunes which are presently vegetated most likely formed at an earlier time (e.g., Pleistocene); vegetation takes place later in the successional development of dunes. With sea-level rise, these dunes erode and serve as a source of sediment. Therefore, structures that protect the shoreline from eroding may cut the supply of sediment. In contrast, along sheltered shorelines where dunes are unvegetated and still being formed, their formation is the result of aeolian movement of sand from beaches. Therefore, structures that maintain the beach also protect dunes. When bulkheads or revetments are placed high in the intertidal, or even in supratidal locations, to limit erosion, the shoreline remains fixed even as the beach erodes away and some aspects of dune habitat will degrade.

Mudflats and Vegetated Communities

Marshes, mangroves, seagrass beds, and macroalgae stands, as well as mudflats, support highly diverse and productive communities of associated animals (See Figure 4-2 and Chapter 1 for description). Therefore, techniques to mitigate shoreline erosion that change the substrate characteristics will lead to changes

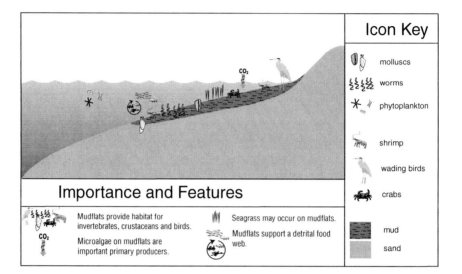

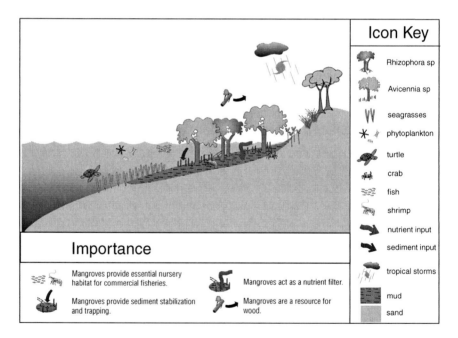

FIGURE 4-2 Conceptual diagram of a mudflat (top), mangrove forest (bottom), and salt marsh (next page) emphasizing their importance along sheltered coasts and outlining processes commonly observed in these communities.

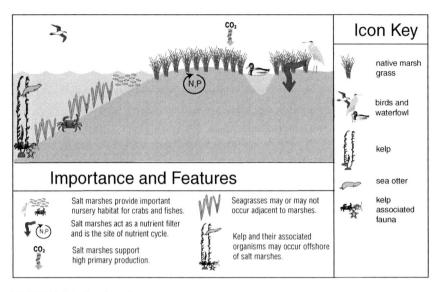

Importance and Features

Salt marshes provide important nursery habitat for crabs and fishes.

Salt marshes act as a nutrient filter and is the site of nutrient cycle.

Salt marshes support high primary production.

Seagrasses may or may not occur adjacent to marshes.

Kelp and their associated organisms may occur offshore of salt marshes.

Icon Key

native marsh grass

birds and waterfowl

kelp

sea otter

kelp associated fauna

FIGURE 4-2 Continued.

in the associated plant and animal communities. Ecosystem services of mudflats and vegetated communities include:

Habitat

Marshes, mangroves, seagrasses and macroalgae serve as habitat for a large diversity of organisms from bacteria to mammals (e.g., sea otters in kelp beds). Often the organisms found in coastal plant communities support large local fisheries, such as the scallop fisheries in New England (Heck et al., 1995) and the crab fisheries in the Chesapeake Bay (Ryer et al., 1990). Mudflats also serve as habitat for a variety of infauna, including molluscs and crustaceans.

Nutrient Uptake

Coastal plant communities and their associated epiphytes (microalgae on their leaf or root surface) are quite effective in removing nutrients from the water column. Excessive nutrients in the water column can lead to increased turbidity due to the proliferation of phytoplankton and cause the loss of benthic vegetation such as seagrasses. Marshes and mangroves remove nutrients from land runoff before it reaches coastal waters, and seagrasses further reduce nutrient availability.

Food Production

Coastal plant communities fuel the food web in shallow waters. Macroalgae are often directly consumed by organisms, such as sea urchins and herbivore fishes. In contrast, marshes, mangroves, and seagrasses serve as an indirect food source in the form of detritus. The detritus particles are not only consumed within the specific plant communities but are also exported to adjacent unvegetated areas, such as mudflats. Mudflats are well known to attract migratory birds due to the food they provide (Zhao et al., 2004).

Wave Attenuation

Intertidal vegetation presents a barrier to water flow, providing some wave attenuation. This can be important for protection of the coastal area as seen during the tsunami of December 2004 in the Indian Ocean. In villages protected by a fringe of mangroves, houses were less damaged than in areas where the mangroves were cut (Badola and Hussain, 2005; Danielsen et al., 2005). It has also been speculated that the impact of Hurricane Katrina on Louisiana would have been reduced if local marshes were still relatively intact. Even when unvegetated, mudflats contribute to wave attenuation via their gentle slope, often over extensive areas.

Sediment Stabilization

Wave attenuation (described above) leads to sediment stabilization and less damage to the area colonized by coastal plant communities during storm events.

Maintenance of Biodiversity

Coastal plant communities host a variety of organisms directly on their leaves and roots. As a result, the diversity in these vegetated systems is much higher than in unvegetated areas. Even so, mudflats host a diverse animal community, especially those associated to the sediments.

Recreation

Marshes and seagrass beds are well known as hunting and fishing grounds, respectively. Ecotours are offered through mangrove forests and scuba diving in kelp forests (macroalgae) is a tourist attraction on the West Coast of the United States. Bird-watchers are also attracted to mudflats where migratory birds feed.

Production of Raw Materials

In developing countries, wood from mangroves is used as a fuel for cooking and marsh plants are a source of roofing material. Macroalgae supply phyco-colloids used in many products, such as ice cream, tooth paste, fertilizers, etc.

The Impact of Shoreline Stabilizing Structures on Mudflats and Vegetated Communities

Surprisingly, little is known about the impact of erosion control structures on adjacent natural communities. The exception is the protective effect of break-waters and sills on marshes. If plant communities are allowed to erode, most of the ecosystem services they provide will be lost.

Marshes

Natural coastal wetlands provide a broad range of ecosystem services. Restored wetlands, especially in areas where the substrate and seed bank are intact, can regain the lost ecosystem services. The restoration may take from 1 to 5 years to reach ecosystem services comparable to a natural or reference system and some may never become fully functional (NRC, 2001). Managing the hydrological function (i.e., sediment elevation and drainage) and the wave climate (possibly with a nearshore structure such as a sill or breakwater) is essential in wetland creation. Once these habitat criteria have been met, marsh vegetation can be established in 2 to 3 years from planting, although sediment characteristics will take longer (perhaps decades) to achieve comparability with natural or reference systems.

Relatively calm conditions are required for marshes to thrive. Therefore, marshes can benefit from breakwaters and sills. Quite often marshes are planted in combination with the installation of these structures to provide sediment stabilization. The success of such hybrid projects (structure-marsh hybrid) depends on the elevation of the marsh habitat because the plants cannot tolerate excessive periods of inundation. Over longer periods of time (possibly decades) bulkheads and seawalls have the potential to be detrimental to marshes, especially in areas with rapid relative sea-level rise. Bulkheads and seawalls are usually placed landward of the marsh preventing migration of the marsh shoreward as land becomes inundated, resulting in loss of the marsh as water levels rise or the marsh edge erodes. Bulkheads, revetments, groins, bridges, culverts, and diking, restrict tidal flow in marshes. Restricting flow leads to freshening of the marsh, increasing the likelihood for *Phragmites* invasion and loss of naturally occurring marsh grasses that support ecosystem functions. The impacts of groins and revet-ments on marshes have not yet been thoroughly studied and documented.

Mangroves

For mangroves, establishment takes longer than for marshes (3 to 5 years, see Lewis et al., 2005). The ecosystem services provided by these restored or created habitats include all of those listed for natural systems, but it takes time for the systems to mature. Algal production dominates at first, followed by grasses. Fish are likely to begin to use these areas as refuges almost immediately and gradually for foraging. The benthic infauna are slower to stabilize because the sediment conditions may take several years to evolve. Birds and mammals, depending upon habitat requirements, may be present during restoration and creation or immediately thereafter.

Just like marshes, mangroves can benefit from breakwaters and sills as they also need relatively calm conditions to thrive (Markley et al., 1992; Field, 1997; Snedaker and Biber, 1997; Milano, 1999). The impact of bulkheads, seawalls, groins and revetment on mangroves has not yet been documented.

Macroalgae

Breakwaters, sills, bulkheads, and seawalls are built with rocks, concrete or wood which are suitable substrates for macroalgae. As a result, the installation of these structures can lead to the growth of macroalgae on the subtidal portion of the structures if sufficient light is available. The algal diversity on these structures is usually lower than that found on rocky shores and new, invasive species can spread relatively quickly (see case study mentioned later in this chapter).

Seagrasses

Except for one study along a highly exposed area (2 meter waves) in Japan (Dan et al., 1998), the effect of breakwaters and sills on seagrasses has not yet been quantified. In this study, seagrasses grew in an area protected by a breakwater while they were absent in the adjacent unprotected area. The authors conclude that the breakwater is beneficial to the seagrasses along this wave-exposed shoreline. In the Chesapeake Bay at a relatively low energy environment (0.4 m storm waves), it appears that breakwaters can be detrimental to seagrasses over long periods of time (decades). Breakwaters tend to accumulate organic and fine particles shoreward of the structure, slowly making the area unsuitable for seagrass growth (Koch, 2005).

Although no peer-reviewed studies were found in a literature survey, it appears that seawalls and bulkheads can be detrimental to seagrasses when trees are planted at the edge of the structure. Trees tend to shade seagrasses in the shallow waters adjacent to the structures. Also, as waters adjacent to revetments and seawalls become deeper, there may not be sufficient light to support seagrass growth, leading to the complete loss of this vegetation type. This remains to be quantified.

Bluffs

Ecosystem Services

The ecology of marine bluffs (see Figure 4-3 and Chapter 1 for description) has been less thoroughly studied than other shorelines. As a result, ecosystem services have not been adequately documented, but are likely to include:

Habitat

Bluff habitat varies depending on differences in substrates, rainfall, wind exposure, and other physical factors. Bluffs, particularly forested bluffs, provide many unique and important habitat features. They tend to be covered in a variety of tree and groundcover species, with stable bluffs supporting old-growth forests and unstable bluffs being suitable for early successional species. These forest habitats provide cover for a variety of terrestrial organisms; in particular, they provide secure nesting sites and hunting perches for eagles, ospreys, herons, and a variety of other avifauna. Unvegetated and rocky bluffs can also provide habitat for cavity- and ledge-dwelling birds such as cliff swallows and peregrine falcons. An added benefit of some of these steep habitats is that they are relatively difficult to access, reducing human disturbance. These habitats are particularly significant

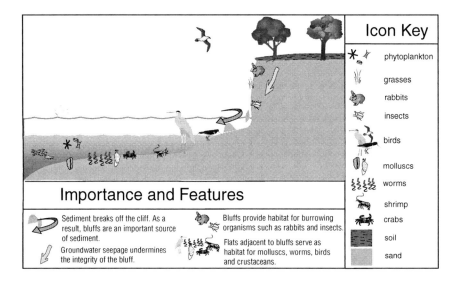

FIGURE 4-3 Conceptual diagram of a bluff emphasizing its importance along sheltered coasts and outlining processes that occur on and adjacent to cliffs.

because many of the resident species are threatened or endangered. Large organic debris (e.g., fallen debris), tree leaves, and marine plant debris, tend to drift and accumulate along banks and bluffs. This substrate provides a complex habitat for insects, amphipods, and other organisms. Another potentially unique habitat associated with bluffs are groundwater seeps; while it is hypothesized that these seeps may host unique or rare species, they are relatively unstudied (Thom et al., 1994).

Nutrients and Groundwater Dynamics

Groundwater flow and associated nutrients, while poorly understood, may be an important service provided by bluffs. Bluff geomorphology will greatly affect the flow of groundwater to surrounding habitats such as beaches and marshes. This affects the delivery rate of nutrient inputs, such as nitrogen and phosphorus, by groundwater to these habitats (Thom et al., 1994).

Shading

An important but potentially overlooked service provided by vegetated bluffs is shading. Overhanging vegetation can shade beach or marsh substrates and near-shore waters, reducing and regulating temperatures. This can have a significant affect on species living in the shaded habitat. For example, shading may be crucial to juvenile salmonids in Puget Sound, which rely on the overhanging vegetation to maintain water temperatures, provide protective cover, and supply detritus and habitat for terrestrial prey items such as insects (Levings et al., 1991).

Food Production

Primary and secondary production on bluffs will depend on the type of habitat, but vegetated bluffs in particular provide food for a variety of organisms, including insects, small mammals, and raptors.

Recreation

Unvegetated bluffs can offer unimpeded views of the sea, making them appealing for housing development. Some bluffs may be attractive for recreation because they provide a means of access to the beach, while steeper bluffs may be attractive because they offer a physical challenge to climbers. The unique habitat provided by bluffs may draw bird-watchers and other naturalists.

Wave Attenuation

Bluffs are very effective at wave attenuation although excessive wave energy leads to erosion at the toe of the bluff.

Sediment Source

Bluffs are a source of sediment for beaches, marshes, and mudflats; changing bluff erosion rates will also alter accretion rates in these depositional systems.

Maintenance of Biodiversity

As previously mentioned, bluffs can be important nesting and hunting habitat for threatened and endangered bird species such as bald eagles and peregrine falcons. Also, old-growth forests, groundwater seeps, and large organic debris drifts provide unique and often complex habitats that are important elements influencing diversity.

Production of Raw Materials

Bluffs may be mined/quarried for their raw material—sand, gravel, stone, and rock. Forested bluffs may also be a source of timber.

The Impact of Shoreline Stabilizing Structures on Bluffs

One of the major impacts of bluff stabilizing structures will be on the services provided by bluffs to downstream ecosystems, such as beaches and marshes. Bluffs are a source of groundwater, nutrients, sediments, and organic debris (i.e., leaves and trees). Structures may affect the natural groundwater regime and drainage of the bluff habitat, thus impacting physical processes such as slope stability. Modifications to the bluff morphology and vegetation associated with stabilizing structures will affect the species that rely on those habitats, particularly sensitive and essential nesting habitat for birds. Shoreline structures may also decrease overhanging trees and shrubs, thus decreasing the shading of substrates below the bluff and affecting species that rely on this shading.

The integrity of sedimentary bluffs depends on the health of the adjacent beach and the capacity of the beach to attenuate waves. Therefore, any negative impact that shore-stabilizing structures impose on beaches (see section on the "impact of shoreline stabilizing structures on beaches and dunes") will also affect bluffs.

ECOSYSTEM SERVICES PROVIDED BY TECHNIQUES TO MITIGATE SHORELINE EROSION

Summary

Shoreline stabilizing structures also provide some ecosystem services. Although usually different and of a lower quality than those of the natural environment, the ecosystem services provided by shoreline stabilizing structures should be considered in the design and implementation of projects to protect eroding areas. The challenge for shoreline planners and permit agencies is evaluating the postconstruction services relative to those in place prior to erosion mitigation. In some instances, specific services provided by the mitigation structure or approach are desirable to specific interest groups, e.g., habitat for specific recreational fishing species, making the consideration of the change in services more complex. A summary of ecosystem services provided by the major types of shoreline protection structures is presented below and presented in Table 4-3.

Bulkhead

The location of the bulkhead is essential in determining its effect on its environment. Spalding and Jackson (2001) show that bulkheads located high in the intertidal did not affect the beach habitat while a bulkhead in the subtidal promoted the loss of sediment and associated meiofauna. When partially submersed, bulkheads could serve as a substrate to molluscs, algae, and associated organisms. Therefore, partially submersed bulkheads appear to only add a minor ecosystem service (if any at all) to an area. The losses of ecosystem services overweigh the gains.

Revetment

Positive and negative effects have been reported for revetments (Quigley and Harper, 2004). Negative effects were observed when revetments led to the loss of natural vegetation which was shown to always be superior in providing ecosystem services than the engineered structure. Positive effects were observed when revetments were installed in combination with other environmental management practices, such as storm water management, in highly degraded areas. The stones used in revetments become colonized by micro and macroalgae which provides food for organisms. Additionally, the rocks serve as habitat for filter feeders, and these can improve water clarity by removing particulates, including microalgae (Newell and Ott, 1999). Bryophytes and associated meiofauna on riprap have also been found to increase spatial diversity (Linhart et al., 2002). Sculpins were more abundant at revetments than sandy or cobble beaches (Toft et al., 2004). In summary, revetments may provide some ecosystem services (mainly habitat) particularly in areas undergoing extensive erosion. However,

TABLE 4-3 Summary of Ecosystem Services Provided by Natural Coastal Ecosystems as Well as by Commonly Used Techniques to Mitigate Shoreline Erosion

Ecosystem Services	Natural Coastal Ecosystems						Techniques to Mitigate Shoreline Erosion					
							Harden shoreline				Trap or add sand	Plant it
	Sandy beaches	Sand dunes	Mudflats (microalgae)	Marshes and mangroves	Seagrasses and macroalgae	Bluffs	Bulkheads and seawalls	Revetments	Groins	Breakwaters and sills	Beach nourishment	Marsh/mangrove planting
Habitat *fishes*	○	○	◉	●	●	○	○	◉	○	●	◉	●
mollusks	●	○	●	●	●	○	◉	◉	◉	◉	◉	●
crustaceans	◉	○	◉	●	●	○	–	○	○	○	○	●
turtles	●	◉	–	○	◉	○	–	–	–	–		○
birds	◉	●	●	●	○	●	–			~	◉	●
Nutrient uptake/cycle	◉	○	◉	●	●	○	○	○	○	○	○	●
Food production	○	○	◉	●	●	◉	○	○	○	○	○	●
Wave attenuation	◉	●	○	◉	◉	●	○	◉	◉	●	○	◉
Sediment stabilization	–	●	◉	●	●	–			◉	●		●
Gas regulation	○	–	○	●	●	–	–	–	–	–	–	●
Biodiversity	◉	●	◉	●	●	●	○	○	○	○	○	●
Recreation	●	●	○	●	●	◉	◉	○	◉	◉	●	●
Raw materials	●	●	○	●	●	●	○	○	○	○	○	●
Aesthetic value	●	●	◉	●	●	●	–	–	–	–	●	●

NOTES: The darker symbol (●) represents the highest degree of contribution for the ecosystem service listed while the lighter symbol (○) represents a low contribution. The ◉ symbol represents an intermediate contribution and a dash (–) suggests that the ecosystem service in not relevant or nonexistent. Please note that the ecosystem services assigned to techniques to mitigate shoreline erosion are best estimates. Extensive research is still needed to determine the ecosystem services provided by these techniques.

even in these degraded environments, some ecosystem services will be lost with the installation of a revetment.

Breakwater and Sill

Coastal vegetation that thrives in sheltered environments can benefit from breakwaters, at least in the short term. Kelp (*Macrocystis pyrifera*), marsh (*Spartina alterniflora*), and seagrasses (*Zostera marina*) have been shown to benefit from breakwaters along high energy shorelines (Rice et al., 1989; Allen et al., 1990; Rennie, 1990). Some caution is needed though, as it appears that the accumulation of fine and organic particles shoreward of breakwaters may lead to the loss of seagrasses in the long term (Koch, unpublished data).

Breakwater units, if built of stone, have also been shown to provide hard substrate that is beneficial to algae, barnacles, and oysters, and creates a foraging area for fish (USACE et al., 1990). In one estimate, the primary production of algae colonizing the hard substrates of a marina (breakwaters, pilings) apparently compensated for the loss of primary production due to deepening of the nearshore zone during construction (Iannuzzi et al., 1996). The benefit of breakwaters can be enhanced when these are covered with oysters which provide the ecosystem service of water filtration and improvement of water quality (Newell and Koch, 2004).

It has been proposed that breakwaters can provide a variety of benefits, particularly to fish communities. However, the evidence for these benefits is mixed. Some studies suggest that breakwaters can serve as a habitat for fishes (Stephens et al., 1994) and often have a higher fish species richness than natural reefs (Lincoln Smith et al., 1994). As a result, breakwaters could contribute to the fish larval pool (Stephens and Pondella, 2002). In contrast, another recent study suggested that shoreline hardening will have a negative impact on certain fish species and their nearshore habitat (Seitz, R.D., et al., 2005). Breakwaters have also been shown to have lower observed total diversity (including all plants and animals) than that of rocky shores (Moschellaa et al., 2005). While it is not valid to compare breakwaters to rocky or sandy beach habitat, clearly there will be a trade-off when a natural habitat is replaced by a man-made structure.

When an eroding bank, narrow beach, or nearshore is changed to a stable bank, marsh or stone breakwater or sill, there is a trade-off that affects habitats. The encroachment should only be as much as the level of protection required. The stable marsh fringe and stone provide a myriad of desirable habitats, but at the expense of a narrow band of benthic communities, requiring an evaluation of the trade-offs by state regulatory agencies similar to those conducted for proposals to add or trap sand. Impacts to adjacent, unprotected coasts must also be considered.

In summary, breakwaters may provide some ecosystem services related to fish, oyster, and plant habitat, and may contribute to primary production although to a limited degree. Breakwaters may also increase spatial biodiversity although this ecosystem service may not always be welcomed (i.e., introduction of new species to a system) but, when compared to natural systems such as rocky shores, biodiversity decreases.

Beach Nourishment

The most detrimental direct effect of beach nourishment is the burial of shallow reefs and invertebrates reducing food availability for birds, fishes, and crabs (Peterson and Bishop, 2005). There is basically a replacement of habitat from nearshore benthic community to an intertidal and supratidal beach and dune. This is not always viewed as positive and some states are reluctant of allow this option even though it is a nonstructural alternative. The benefits of beach nourishment have not yet been appropriately quantified (Peterson and Bishop, 2005). The addition of vegetation in the form of dune grass plantings is recommended not only as added level of protection but an important habitat component. Rarely considered but also of potential importance are habitat changes in the borrow area for the beach fill (Nordstrom, 2005) where deep holes are created in the nearshore and benthic habitats are dramatically changed, at least immediately after dredging.

CUMULATIVE AND SECONDARY IMPACTS OF TECHNIQUES TO MITIGATE SHORELINE EROSION

As in many instances of mitigation (NRC, 2001), the cumulative effects of multiple shoreline structures or protection measures are rarely addressed. The scale of cumulative erosion control measures in some areas is massive. One of the best-documented examples is Mobile Bay, where by 1997, 30 percent of the bay's shoreline was armored (Douglass and Pickel, 1999). This has led to the loss of 4-8 hectare (approx. 10-20 acres) of intertidal area and 6-13 kilometers (approx. 4-8 miles) of shoreline. The loss of intertidal habitat as a result of extensive placement of vertical bulkheads certainly limits recreational opportunities for local residents and reduces the availabilities of shallow intertidal habitat for nekton, although any population level effects of this loss of habitat are difficult to quantify. Similar assessments in Puget Sound (Gabriel and Terich, 2005) showed that shore protection structures have proliferated over the last 40 years with currently over 80 percent of properties protected from erosion in some areas. It is the goal of this chapter to provide factual information, not to evaluate the negative or positive impacts of the changes.The placement of mitigation structures in separate places along a shoreline can result in disruption of linkages between environments. This is of ecological concern as it can potentially interfere both with the coupling of systems in terms of energy and nutrient exchange, and as habitats of common use by organisms are disconnected. The importance of linkages between subtidal and intertidal marsh habitats for nekton has been well established (Weisberg and Lotrich, 1982). However, the effects of cumulative changes in shoreline character on system level provision of ecosystem services, such as nursery habitat for nekton, is difficult to assess as the specific value of the habitats is itself poorly defined (Beck et al., 2003).

A recent review by Airoldi et al. (2005), showed that the proliferation of shore protecting structures can have critical effects on regional species diversity. A large number of nearby structures can disrupt natural barriers (e.g., extensive sand beaches serve as a barrier for the dispersion of organisms associated with rocky shores) enhancing the dispersal of species characteristic of rocky shores in regions that were naturally poorly connected. This can also lead to dispersal routes for invasive species as seen in northeastern Italy where the invasive macroalga *Codium fragile* subsp. *tomentosoides* rapidly expanded along the 300 km (approx. 190 miles) of protected shorelines. In summary, a landscape approach that includes population dynamics needs to be considered when permitting shore protecting structures. The best examples on how to manage natural resources at a regional scale can be found in the design of marine protected areas (see Kinlan and Gaines, 2003; Lubchenco et al., 2003), which also needs to consider habitat fragmentation, loss of habitat connectivity, and dispersal of species at the landscape level (Airoldi et al., 2005).

CASE STUDIES

Ecological Impacts of Low-Crested Breakwaters in Europe

Extensive work on the ecological consequences of building low-crested structures (LCS; i.e., sills) has been recently completed in Europe (see Moschellaa et al., 2005). Six structures exposed to a variety of hydrodynamic conditions were studied in detail. Surprisingly, despite geographical, hydrodynamic, and engineering differences, some clear patterns emerged. Ecological changes induced by the LCS were mainly a result of a modification of hydrodynamic conditions and sediment composition landward of the structures. The presence of sills increased the overall species diversity of infauna mainly due to a larger number of species found in the sheltered area (Martin et al., 2005). The introduction of new species is seen as a negative transformation of the environment by the authors (Martin et al., 2005). The sills also attracted fish species typical of rocky shores, but these remained in the juvenile stages. As a result, intense fisheries activities around the structures were absent. Additionally, the installation of sills or breakwaters killed the benthic organisms in the footprint of the structure. The accumulation of fine and organic rich sediments landward of the LCS, especially those along relatively sheltered coastlines, resulted in a shift in the composition of the infauna (Martin et al., 2005). In some instances, the same processes that led to the accumulation of fine and organic particles also led to the excessive accumulation of macroalgae such as *Ulva lactuca* creating unpleasant conditions (rotting algae) for recreational activities.

In order to minimize the negative impacts associated with low-crested breakwaters (altered hydrodynamics and deposition of fine and organic sediment),

Martin et al. (2005) suggest that the structure be built: (1) as far away from shore as possible, (2) as porous as possible, (3) with as much overtopping (i.e., water flowing over the structure) as possible, (4) with maximum gap size and number, (5) without beach nourishment, (6) without lateral groins, and (7) be avoided if at all possible, in areas dominated by fine sediments. The manuscript concludes that "the number of LCS should be reduced to the minimum necessary to protect the coast, avoiding large-scale effects of habitat loss, fragmentation and community changes." These ecological considerations have been incorporated into a model to design more environmentally friendly structures to protect shorelines from erosion (see Zanuttigh et al., 2005).

Mills Island, Chincoteague Bay, Maryland

In some locations, shoreline erosion creates habitat for ecologically important species. Shoreline stabilization in these areas would lead to loss of habitats and ecosystem services. This is the case at Mills Island, located on the western shore of Chincoteague Bay, MD, an area of high relative sea-level rise. This marsh island is eroding at a rate of half a meter (approx. 2 ft) yr^{-1} exposing compacted peat in the subtidal areas. This sediment is not suitable to seagrass growth, a plant type that serves as habitat for a variety of commercially and ecologically important species. As a result, marsh erosion is also leading to seagrass loss with the exception of areas where dunes located within the marsh island are also eroding (Wicks, 2005). Where the dunes are eroding, a layer of sand is deposited on top of the compacted marsh peat making it a suitable seagrass substrate. The construction of any structure to reduce or stop shoreline erosion will lead to deepening of the offshore area and will no longer allow sand to cover the unsuitable compacted marsh peat, leading to the loss of seagrass habitat (Wicks, 2005). Therefore, the benefits of shoreline erosion (e.g., sediment supply) need to be considered when developing a regional shoreline protection plan.

FINDINGS

- A general lack of information exists about the ecosystem services provided by structures to mitigate shoreline erosion.
- Techniques to mitigate shoreline erosion lead to the loss of some ecosystem services.
- The loss of ecosystem services associated to the mitigation of eroding shorelines can be localized when only a few structures exist within a system but can alter the whole area (even where such structures are absent) when a certain critical percentage of shoreline modification is exceeded.
- Techniques to mitigate shoreline erosion may contribute some ecosystem services, but not the range of services provided by natural systems.

5

The Existing Decision-Making Process for Shoreline Protection on Sheltered Coasts

Aspects of decision-making and the factors that influence it have been discussed in previous chapters. Similarly, these chapters have detailed the various options that exist to address the problem of eroding shorelines. The focus of those chapters, however, has not been on the *process* of making the decision to protect a shoreline or what steps are necessary to reach implementation of a solution. This chapter presents an overview of the current decision-making process for erosion control on sheltered coastlines. The process includes a discussion of who selects responses for shoreline protection and highlights the key factors that influence and constrain those that make the decisions, as well as the complexities of altering and regulating the shoreline.

DECISION-MAKERS

The property owner who perceives a problem with erosion of the shoreline usually initiates the response to the problem. After observation of a problem, an investigation into possible solutions ensues. If the "do-nothing" alternative is discarded, the property owner must proceed through a regulatory process that will involve other decision-makers and, depending on the size and complexity of the situation, be complex and lengthy.

For any segment of eroding shoreline, the choice of which mitigation option to implement is affected by multiple decision-makers. These decision-makers can be broadly grouped into four classes:

- property owners,
- experts and consultants (such as civil engineers),
- government regulators, permitting and compliance officials, and
- policy-makers or law-makers.

Each of these classes of decision-maker imposes limitations or requirements on the scope of the options that can be considered by the others involved in a particular project. The motivations and constraints of the different groups of decision-makers vary depending on their relation to the property, their knowledge of different types of shoreline protection options, their stewardship responsibilities, their professional interests, regulatory framework, legal precedence, and local preferences.

For example, proponents of shoreline protection are usually property owners driven by a desire to preserve upland area and value or by a desire to protect, create, or restore recreational opportunities that a beach may provide, for example. They seek an outcome that will protect and maximize their uses of the shoreline and their investment. The realities of private property insurance or the National Flood Insurance Program (NFIP) may also influence property owners' decisions. NFIP, run by the Federal Emergency Management Agency (FEMA), informs the zoning decisions made by communities. According to the Coastal Engineering Manual (U.S. Army Corps of Engineers, 2002b): "Any structural or nonstructural change in the design, construction or alteration of a building to reduce damage caused by flooding and flood related factors (storm surges, waves, and erosion) is considered a flood proofing alternative by FEMA," (see Box 5-1).

BOX 5-1
National Flood Insurance Program (NFIP)

The NFIP was established with the passage of the National Flood Insurance Act of 1968 (P.L. 90-448). It is a voluntary program based on an agreement between flood-prone communities and the federal government to reduce future flood risk to new construction in floodplains. The Federal Emergency Management Agency (FEMA) publishes maps, called Flood Insurance Rate Maps (FIRMs), to show areas in the community that have a 1 percent or greater chance of flooding in any given year, known as Special Flood Hazard Areas (SFHAs). FIRMs are developed by engineering companies, other federal agencies, or the community at risk and are reviewed and approved by FEMA. FIRMs are used by property owners, buyers, insurance agents, and lending institutions to locate properties and buildings in flood insurance risk areas. Community officials use the FIRM for their area to administer floodplain management regulations and to mitigate flood damage.

SOURCE: http://www.fema.gov/plan/prevent/fhm/mm_mca.shtm.

Property owners include individuals or groups with large and small private holdings, as well as entities representing all levels of government. Constraints on their decisions include (1) costs, (2) feasibility, (3) regulatory permissibility, (4) length of time to permit, and (5) lack of knowledge about the potential options available to them.

Experts and environmental consultants, including government scientists and engineers, represent the second class of decision-makers. Property owners rely on consultants to provide most of the information that enters their decision-making process. In turn, consultants are motivated by their client's desired outcome. The geomorphology and hydrodynamics of the site present a number of constraints in addition to cost, feasibility of success, ease of permitting, and the consultant's familiarity with various mitigation options. In general, the larger a project, the lengthier, more time-consuming, and more complex design and permitting become.

Government regulators, permitting officials, and compliance officers at all levels of government are driven by the legal mandates and official policies that apply to and direct activities along the shoreline. Regulators also include county and town civil engineers, building inspectors, zoning officials, and local land use planners. The shoreline and adjacent lands are some of the most highly regulated real estate that exists. The purpose of regulations governing the shoreline is generally to protect public trust areas, such as beaches, wetlands, tidelands, and nearshore coastal waters, while at the same time balancing private property rights. The motivation of a government regulator may often conflict with the desires of property owners. Often regulators seek to limit encroachments into public trust areas—generally the area below mean high water line—see Figure 1-1 (in Chapter 1), making permitting of structures located directly adjacent to eroding upland easier and quicker than permitting of structures in or on the nearshore areas. The feasibility of success and knowledge of the proposed shoreline erosion mitigation options are also major factors that influence regulatory decision-making. Public access to or around a proposed mitigation project can also be a major consideration.

Policy-makers comprise the fourth group of decision-makers. These individuals draft and approve federal, state, and local statutes and ordinances that govern activities along the shoreline. Most policy-makers are elected state, county, and municipal officials who must balance public input with the information they receive from respected experts and interest groups. Court cases and legal precedent, such as the evolving nature of the public trust doctrine (see Box 5-2) and wetlands legislation, as well as private property rights and potential takings claims, also influence these decision-makers. These legal issues can limit some shoreline erosion mitigation options, such as extensive fill of wetlands or building restrictions in coastal areas.

BOX 5-2
Public Trust Doctrine

The Public Trust Doctrine (PTD) is a legal concept that often has great impact on how governments manage their tidelands and nearshore waters. The government is the trustee of these areas for the benefit of the public and maintains this stewardship responsibility even though in some cases it may previously have decided to privatize some submerged sovereign lands (see NRC, 1999).

The original public trust interests included navigation, fishing, and commerce (*Martin v. Lessees of Waddell*, 41 U.S. 367 (1842) and *Shivley v. Bowlby*, 152 U.S. 48 (1894)). However, recent case law in some states since the early 1970s has produced an evolution in the interests that the PTD covers. Thus, depending on state law, the PTD interest may extend to open space protection, environmental quality, and recreation interests.

The PTD has implications for all decisions regarding shoreline erosion control options that inevitably produce an impact on public trust lands. The outcome will depend on the dominant interest at stake. For example, some erosion control options, such as wetland creation and "living shorelines" may impede the public trust interest of navigation while enhancing other public interests such as environmental quality and fishery habitat. Other responses, such as bulkheads, may degrade the quality of nearshore environments (e.g., reduce their quality as fish habitat), but maintain navigation. If protecting natural shorelines, wetlands, and beaches is a priority in an area, then some responses to erosion such as vertical walls may not be feasible. In other areas, protection of navigation interests might be paramount and lead to erosion responses that conflict with conservation of natural areas.

An additional issue that often complicates decision-making is public access. Not only does common law recognize the riparian right of access to navigable waters, it also guarantees the public's right to navigate on waters. This later concept may create obstacles for proposals that interfere unreasonably with the public's access to navigable waters, as well as the public navigation interest. Erosion control options, such as beach creation, may also create new opportunities for public access to the fringes of navigable waters. For example, in Mobile Bay, Alabama, property owners are warned by regulators that the construction of beaches—a mitigation option that increases habitat value as compared to a seawall which does not—will require that the public be granted access to the shoreline in places where it may not have had access previously (Douglass, 2005a).

The following table (Table 5-1) summarizes the motivations, information sources and needs, and area of influence of the four categories of decision-makers.

As noted in Table 5-1, the motivations and influences affecting individuals and groups of decision-makers vary. Constraints such as knowledge of different types of shoreline protection options, regulatory precedence, costs, and local preferences are equally important in decision-making and differ among the groups of

TABLE 5-1 Characteristics of Various Groups of Shoreline Protection Decision-Makers

Decision-Maker	Objectives	Information Needs	Information Sources	Area of Influence
Property Owners	• Maximize the use of their property • Aesthetics • Maximum property value	• Effectiveness • Cost • Feasibility	• Handbook and/or online info • Expert/consultant • Government regulator • Neighbors • Flood zone maps	• Individual's property, as well as neighbors' property Note: Large public property owners can have broader geographical influence.
Experts and Consultants (includes government scientists and engineers)	• Satisfy the client • Make a profit • Maintain credibility	• Knowledge of shoreline protection options (Structural and nonstructural) • Feasibility (i.e., ease of permitting) • Physics, geomorphology, and ecology	• Professional networks • Experience • Field work • Trade publications • Government agencies • Vendors • Formal Education	• Geographical region in which they work
Government Regulators, Permitting and Compliance Officials	• Implement and enforce the regulations • Resource stewardship	• Knowledge of shoreline protection options (structural and nonstructural) • Physics and geomorphology, and ecology • Legal mandates • Public trust responsibility • Constraints imposed by other regulatory programs	• Reports or the NRC and other expert bodies • Professional networks • Experience • Consultants • Formal education • Legal counsel	• Jurisdiction in which they work
Policy-makers and Law-makers	• Reelection • Maintaining the tax base • Resource stewardship • Serving their constituents • Environmental quality • Quality of life • Public health, safety, and welfare	• Public trust responsibilities • Current law; its impacts and any unintended consequences • Perception and understanding of the problem to be solved	• Press • Constituents • Staff (trusted experts in the field) • Government agencies • NGOs	• Their jurisdiction, as well as their colleagues' jurisdictions

decision-makers. Ultimately, the final decision regarding selection of a shoreline erosion mitigation approach is guided by the constraints and requirements of the decision-making process.

STRATEGIES AVAILABLE TO ADDRESS THE PROBLEM

This chapter focuses on decision-making to protect eroding shoreline properties. A property owner who wants to protect an eroding shoreline will, with support from a consultant, consider the options discussed in Chapter 3 (shoreline hardening, vegetative responses, addition or trapping of sand). Regulatory authorities have established criteria for these activities that condition their adoption upon satisfaction of the same. The vehicles for this governmental oversight are the permitting systems discussed here. Permitted mitigation activities all tend to be reactive to existing shoreline erosion problems.

One of the constraints mentioned above is the decision-maker's lack of knowledge of shoreline protection options. Most shore erosion studies, both engineering and environmental, have focused on erosion problems on open coasts and as such have placed less emphasis on discussing options for addressing erosion on sheltered coasts. For example, the Coastal Engineering Manual published by USACE (U.S. Army Corps of Engineers, 2002b) devotes 60 pages to the discussion of hard structures used to mitigate erosion (e.g., seawalls, bulkheads, groins, breakwaters, and revetments) but includes only 3 pages on reefs, sills, and wetlands.

Parallel strategies have evolved that attempt to proactively reduce the problem and avoid erosion losses by restricting or prohibiting development in erosion zones ("hazard zones"). The report mentions some of these strategies to manage land use at the beginning of Chapter 3—comprehensive master plans, zoning, acquisition of sensitive or hazardous lands ("property buyouts"), setbacks zones, and subdivision ordinances. Evidence suggests that land use planning techniques are the most effective approaches for promoting sustainable mitigation from hazards and avoiding infrastructure losses (Burby et al., 2000). If communities succeed in limiting or prohibiting construction in shoreline areas that display a high risk of erosion, then shoreline erosion diminishes as a problem.

Analogies exist between mitigation of shoreline erosion and mitigation of flood losses. While requirements to elevate buildings in floodplains may reduce losses, these structural requirements do not limit new construction in floodplains (Holway and Burby, 1993). Similarly, regulation of the type and standards of shoreline protection activities does not directly restrict construction in these areas. In fact, shoreline protection may often be an incentive for further shoreline development, which may increase the exposure to additional losses (Burby et al., 2000).

Zoning and similar proactive land use planning techniques are tools that can successfully limit development in hazardous areas and, therefore, be more effective at reducing property losses. These direct land use approaches that attempt to avoid the problem include acquisition of sensitive and/or hazardous lands, relocation of structures, shoreline setbacks, zoning, and comprehensive planning that incorporates the eroding shoreline.

This report focuses primarily on direct mitigation responses to shoreline erosion—specifically structural, vegetative, and sand addition options. However, we realize that proactive land use planning techniques are essential elements for addressing this problem and avoiding losses to property and infrastructure. Regulation of land use and limitation of development are the central concepts in the avoidance of losses in these hazardous areas. These strategies are generally under the purview of local governments—sometimes with standards and mandates that originate at the state level.

In summary, decision-makers are influenced by:

- financial and ecological costs, both incurred and avoided,
- legal precedent, political feasibility and acceptability, time required for permitting,
- site conditions, durability of the erosion control technique, spatial scale of the project, and
- ancillary benefits such as public access, new revenue sources, ecosystem services provided, or broad public support.

PERMITTING REQUIREMENTS FOR SHORELINE PROTECTION ON SHELTERED COASTS

Federal, state, and/or local governments usually regulate shoreline protection activities by means of a permitting system that establishes criteria and then evaluates whether the proposed action conforms to the accepted criteria. The precise permit requirements depend on the jurisdiction, the exact location of the proposed activities, and, of course, the type and size of the protection strategy proposed. Although great variation exists among shoreline permitting systems in the various jurisdictions in this country, the federal permitting system creates some national uniformity and often influences the permitting frameworks at lower levels of government. The following section discusses the major federal legislation and accompanying state reviews potentially triggered by shoreline protection projects.

Federal Permits

If an activity will occur in waters of the United States, generally anywhere along a sheltered shoreline, then a permit from the U.S. Army Corps of Engineers (USACE) is required.

Two federal laws serve as the basis for the federal regulation of shoreline activities: the Rivers and Harbors Act of 1899 and the Federal Water Pollution Control Act (FWPCA; Clean Water Act) of 1972. Through its administration of both statutory programs, USACE plays the central role in the regulation of shoreline protection projects.

Congress has granted the Secretary of the Army (USACE) the power to regulate navigation and navigable waters through the Rivers and Harbors Act of 1899. Section 10 of this legislation (33 U.S.C. sec. 403) prohibits an individual "to excavate or fill, . . . modify the course, location, or condition . . . of any channel of any navigable water of the United States" without authorization of the USACE. In tidal areas, the geographical extent of "navigable waters of the United States, as used in the Rivers and Harbors Act, extends to all areas covered by the ebb and flow of the tides to the mean high water (MHW) mark in its unobstructed natural state." USACE regulations define the landward limit of this agency's jurisdiction in coastal waters to be the line reached by the MHW even though some of the water body may be very shallow or support wetland vegetation (33 C.F.R. sec. 329.12) (see Figure 1-1). An estimate of the precise line (averaged over the 18.6-year tidal cycle, according to *Borax Consolidated Ltd. v. Los Angeles*, 296 U.S. 10 (1935)) may refer to vegetation lines or to physical markings.

Federal regulatory authority may extend further inland than the MHW line, depending on the geographical circumstances, according to the amendments to the FWPCA that Congress passed in 1972 (Clean Water Act [CWA]). Section 404 (33 U.S.C. sec. 1344) granted the USACE authority to regulate the discharge of dredge and fill material into navigable waters. While section 404 uses the term "navigable waters," later in the legislation, Congress defined these as "waters of the United States" (33 U.S.C. sec. 1362(7)). For the purposes of section 404, the USACE regulations define the landward limit of its jurisdiction in coastal waters to be the high tide line, or in cases where adjacent nontidal waters of the United States are present, the jurisdictional limits of those nontidal waters (33 CFR sec. 328.4(b)).

Both the USACE and the Environmental Protection Agency (EPA) share administrative responsibility for section 404. In most cases, the USACE determines the geographic extent of section 404, but in special circumstances, the EPA can make the final determination on the geographical reach of section 404. The USACE handles the permitting process. Although both permit programs (Rivers and Harbors Act sec. 10 and Clean Water Act sec. 404) may differ in their geographical coverage, the permitting procedures and forms used for both are identical.

The EPA and the USACE regulations both define "waters of the United States" broadly. Not only do they cover traditional navigable waters (i.e., Rivers and Harbors Act), but also interstate wetlands and wetlands adjacent to navigable waters (33 C.F.R. sec. 328.3 - USACE; 40 C.F.R. secs. 112.2 & 116.3 - USEPA). The exact definition of wetlands and their delimitation remain controversial and are based on scientific criteria, as well as politics and social values (40 C.F.R. sec. 230.41; Kalo et al., 2002) For additional information, see also NRC (1995b).

The USACE (2001) issues two types of permits: General Permits (Nationwide and Regional) and Individual Permits (33 C.F.R. sec. 325.5). *General Permits*, in many cases, do not involve individual review of proposed activities and provide expedited authorizations for certain classes of activities that the USACE has determined are similar in nature and cause only minimal individual and cumulative environmental impacts (33 C.F.R. secs. 322.2(f) & 323.2(h)). However, for certain activities General Permits require project proponents to notify the USACE and obtain confirmation that the proposed work is authorized by General Permit, before beginning the work (33 CFR sec. 330.6(a)). General Permits may be nationwide or regional in scope, and their adoption involves normal rule-making procedures, such as public notice of the proposed rule and the opportunity for public comment and perhaps public hearings.

One example of a Nationwide General Permit that may, under certain circumstances, be used for some shoreline protection projects is Nationwide Permit 27 (NWP 27) for Stream and Wetland Restoration Activities (67 Federal Register 2020, 2082-2083, (15 January 2002)). This permit allows activities in waters of the United States associated with enhancement of degraded tidal and nontidal wetland and riparian areas, as well as the creation of wetlands and riparian areas. These activities must not convert tidal waters to other uses, and the use native species of flora for the activity is encouraged. In coastal areas, NWP 27 is used primarily for wetland restoration activities, creation of small nesting islands, and construction of oyster habitat. However, in certain situations, NWP 27 activities could also assist shoreline protection and be part of a vegetation mitigation response to shoreline erosion.

Another general permit, NWP 13 for Bank Stabilization activities, authorizes the construction of structures and fills necessary for erosion prevention (67 Federal Register 2020, 2080). Under NWP 13, the permittee must notify the USACE before beginning the work if the structure is longer than 500 linear feet or uses more than one cubic yard of fill material per running foot placed along the bank below the plane of Ordinary High Water or the High Tide Line. Thus, small bank stabilization activities (less than 500 linear feet in length and using less than one cubic yard of material per foot) can be constructed without notifying the USACE. The NWP 13 does not authorize stabilization projects in wetlands or in special aquatic sites.

Despite the existence of these Nationwide General Permits, they do not have universal application because states can impose conditions that are more restrictive than those of the USACE. These more restrictive regional and state conditions often center on concerns regarding water quality (33 U.S.C. sec. 1341) and consistency with a state's approved coastal zone management plan (16 U.S.C. sec. 1456(c)). The nationwide permits, therefore, ease the permitting process and shorten the approval time for activities like installing bulkheads or other vertical shore protection directly adjacent to eroding upland shorelines.

An Individual Permit is required if the activity does not fall under the conditions of the Nationwide or Regional General Permits. For example, if a property owner wishes to protect an eroding marsh or install a stabilization alternative, such as a sill or breakwater to protect eroding upland, then neither of the nationwide permits mentioned previously could be utilized. The permit applicant would be required to go through the lengthier and more complex individual permit process described here.

These Individual Permits usually have a 30-day comment period during which all interested parties, adjacent property owners, and federal, state, and often local agencies can review the project proposal and submit comments before the USACE makes its decision. Under an abbreviated procedure, the District Engineer may grant a Letter of Permission for some projects determined not to have significant individual and cumulative impacts on environmental values and would have no appreciable opposition (33 C.F.R. secs. 325.2(e)(1) & 325.5(b)(2)). This abbreviated procedure proceeds without the publication of an individual public notice, but with coordination with fish and wildlife agencies and adjacent property owners.

The individual permit application is evaluated by means of a Public Interest Review (33 C.F.R. sec. 320.4), which is used to determine if the activity is contrary to the public interest. This review considers all probable impacts, including cumulative impacts, and considers a broad spectrum of factors that are applied in a case-specific manner during the assessment of the project's benefits and adverse impacts. Factors in the review include conservation, economics, aesthetics, environmental concerns, wetlands, fish and wildlife values, flood hazards, land use, navigation, shore erosion, recreation, water quality, safety, and considerations of property ownership. The Public Interest review process considers (1) general criteria of the public and private need for the work, (2) the practicality of using reasonable alternative locations and methods to accomplish the objective of the proposed work, and (3) the extent and permanence of the benefits and adverse effects on public and private uses. USACE regulations state that unnecessary alteration and destruction of wetlands should be discouraged as contrary to the public interest (33 C.F.R. sec. 320.4(b)(1)). The USACE endeavors to make a decision on an individual permit within 60 days after receipt of a complete application (33 CFR sec. 325.2(d)(3)).

USACE regulations recognize the general right of a property owner to protect against erosion (33 C.F.R. sec. 320.4(g)(2)). While building protective structures is generally allowed, and the USACE minimizes to the extent feasible any adverse impacts to adjacent property, public health and safety, wetland values, and the public interest, the Corps does not usually recommend specific design solutions or alternative measures to vertical stabilization at the erosive face of the shoreline. The regulations limit the district engineer's authority to recommend specific design solutions, to minimize the federal government's legal liability. Landowners often contract with private engineering or environmental consulting firms for advice and design services for bank stabilization activities. In addition, the USACE regulations stress the riparian owner's right of access to navigable waters, as well as the public's right of navigation on waters. Permit applications will generally be denied for proposals that create unreasonable interference with access to or use of navigable waters.

A Federal Permit granted by the USACE for shoreline protection activities may also trigger additional federal regulatory requirements, as well as tribal coordination, depending on the unique circumstances of the case. The following federal statutes may require additional regulatory reviews:

- National Environmental Policy Act (Environmental Impact Statements, Environmental Assessments)
- Coastal Zone Management Act (consistency requirements)
- Clean Water Act (Section 401 Water Quality Certification)
- Fish and Wildlife Coordination Act
- Endangered Species Act
- National Historic Preservation Act
- Sustainable Fisheries Act

A more detailed description of potential federal requirements appears in Appendix D.

State and Local Permits for Shoreline Protection Activities

State Permits

In addition to federal permits, most shoreline erosion control projects require permits based on state wetland, coastal, or shoreline management legislation. State shoreline and land use planning regulations (coastal construction zones, construction setback zones, erosion control easements, special management areas, among others) have their foundations in the state's responsibility to protect the health, welfare, and safety of its citizens. Additionally, the state's proprietary interest in tidelands, generally landward to the MHW line, although a minority of states only own to the MLW line (see Figure 1-1), is the basis for its marine stewardship responsibilities. Moreover, through the common law, the state also has an affirmative duty to protect and promote the Public Trust interests (navigation, fishing, commerce, recreation, open space, and environmental quality) in its submerged lands and tidewaters, even when the state has opted to convey some of its state-owned tidelands or shoreland to private interests and riparian owners. Courts in some states have even held that the state's public trust responsibilities extend to areas landward of the MHW line, such as the dry sand beach and dunes. An additional source of state authority is delegation through federal legislation. For example, state permit approvals are also often triggered by an application for a federal permit and include Clean Water Act Section 401 water quality certification and Coastal Zone Management Act consistency certification.

State coastal management plans developed and approved under the authority of the 1972 Coastal Zone Management Act (CZMA) may also provide the regulatory basis for state permits in shoreline areas. The CZMA provides incentives and national guidelines for states to develop their own coastal management plans tailored to their priority issues and realities. NOAA's Office of Ocean and Coastal Resources Management must approve the state plan before federal incentives begin. The 34 approved state and territorial coastal management plans (out of 35 eligible states and territories) vary greatly in their organizational and institutional approaches, priority issues, definitions of "coastal zone", and strategies adopted to address the problems. Some states, such as North Carolina and California, have created new state agencies possessing regulatory authority—the Coastal Resource Commission or the Coastal Commission, respectively—to oversee the state coastal management program and the establish state standards for coastal activities. Both states' coastal permitting requirements originate from their state coastal commissions. Other states, such as Florida, have opted for different approaches and attempted to foster coordination ("networking") among existing institutions and programs without the creation of a new agency. Regardless of the precise institutional approach, numerous state plans have created and/or incorporated special area management plans and coastal setback zones to protect vulnerable ecosystems, avoid property losses from shoreline erosion, and better manage coastal development. Activities to mitigate shoreline erosion in these areas are strictly regulated and may require state permits.

Some states, such as Oregon, directly engage in land use planning. Oregon integrates coastal management into a state land use planning program that lays out a series of goals. Actions of state agencies, as well as local land use plans, must be consistent with these statewide goals (Beatley et al., 2002). Most states, however, delegate this responsibility of land use planning to local governments (municipalities and counties) without direct state oversight.

Some state and regional efforts have built direct links to local land use plans. For example, Virginia's Chesapeake Bay Preservation Act mandates that all local governments amend building codes, subdivision ordinances, and zoning codes to add specific regulations for wetland, coastal, and bay protection. The Virginia Department of Conservation and Recreation (VA DCR) was charged with implementation of these requirements. Since the tidewater localities were required to adopt and implement the Bay Act, the VA DCR provided electronic versions of the approved resolutions for amending these ordinances.Subsequently, many counties and towns simply adopted the regulations that the State had furnished.

A serious constraint on all these state and local land use efforts to restrict coastal development is the 5th and 14th Amendments of the U.S. Constitution and private property rights. U.S. Supreme Court cases in the past 20 years (*Nollan v. California Coastal Commission*, 483 U.S. 825 (1987); *Lucas v. South Carolina*

Coastal Council, 505 U.S. 1003 (1992); *Palazzolo v. Rhode Island*, 533 U.S. 606 (2001)) have dampened state efforts somewhat. The result has been a softening of regulation, provisions for variances, and perhaps an increase in the use of economic and fiscal incentives, rather than "command and control" regulatory approaches.

Local Permits

Local (county or municipal) land use planning measures and building/ construction permits may pose additional regulatory requirements on the coastal landowner. Local governments' land use policies also exert a great influence over shoreline protection activities, adding another layer of complexity to federal and state requirements that this report has already mentioned. Local government can play a critical role in growth management and hazard mitigation through land use planning strategies. However, some land use approaches considered in this publication that minimize the impacts of shoreline erosion, such as addition of sand, shoreline structures, or even vegetation responses, may actually promote increased shoreline development and, therefore, increase the vulnerability to loss (NRC, 1999).

Local governments possess some clear advantages in addressing shoreline erosion. They are more familiar with the local problems/needs and socioeconomic issues and have access to on-the-ground information. They are often in an excellent position to develop cooperative enforcement strategies with local developers and property owners that can improve the effectiveness of regulations (Burby and Paterson, 1993). Local governments are well-positioned to engage in a consensus-building process in land use planning (Burby et al., 2000). At the same time, local governments often face limitations in managing land use. Their staffs may be small or lack adequate skills; local budgets may be insufficient for regulating development pressures in highly desired areas; and planning boards may be forced to grant variances to avoid takings/inverse condemnation challenges from property owners.

Comprehensive planning and zoning are two key techniques that local governments may select to address shoreline erosion issues (among others). A master plan or comprehensive plan lays out a vision for the direction in which the community wishes to evolve in a period of 10 or 20 years considering the physical and environmental features, hazard zones, development patterns, transportation issues, and social benefits. Zoning establishes a spatial distribution of allowed land uses and their intensities in the community, thus separating land uses. Environmentally sensitive areas and shoreline erosion zones may overlay development zones to generate special zones where development should be restricted or prohibited. Zoning plans must be translated into enforceable zoning ordinances that are respected. This integration of shoreline erosion zones (and other hazard zones, as well) into land use planning is the key element in avoiding infrastructure losses and creating sustainable communities. Both Florida and North Carolina require local governments to prepare local land use plans that include special plans for identified hazardous coastal areas. Both states obligate local governments to adopt minimum state standards (Beatley et al., 2002). For example, under the North Carolina Coastal Area Management Act, the Coastal Resource Commission has set state standards that coastal counties must satisfy when they adopt their required county land use plans. Municipalities also must follow these standards (Beatley et al., 2002).

Local governments may also adopt additional techniques to guide the siting, type, density, quality, environmental surroundings, and growth rate of development. For example, local building codes, subdivision requirements, setback regulations, stormwater management regulations, soil erosion regulations, and vegetative buffer requirements are all tools that may also strongly influence development. Indeed, many of these tools may be adopted and integrated into broader comprehensive plans and zoning approaches.

Despite the importance of good land use planning to avoid losses to shoreline erosion, many sheltered shorelines are already highly developed areas. Thus, the application of reactive strategies (addition of sand, hard structures, and vegetative responses) to avoid property losses is highly relevant in these situations. In these situations, it is too late to consider avoidance strategies by themselves.

Intergovernmental Coordination

The multiplicity of erosion control strategies existing at the various levels of government argues for effective intergovernmental coordination. The permitting process provides some relevant examples. Permitting processes are usually tedious, very time consuming and expensive. Public notices are required; detailed project plans and specifications must be prepared; information on compliance with the other federal laws must be supplied; and consultation with numerous federal and state agencies must occur. Attempts have been made at the federal, regional, and state levels to simplify and streamline various segments of the permitting process. Examples of this include issuance of Nationwide Permits by the USACE and of general permits and certifications by the states, as well as implementation of joint public notice processes that combine portions of the permit review process of several state and federal agencies.

Intergovernmental coordination mechanisms often exist to facilitate the evaluation of these multiple permits. In some cases, the USACE grants State Programmatic General Permits when the state regulatory program is similar to the USACE program. This attempts to avoid duplication by streamlining the permitting process. Programmatic General Permits are based on existing federal, state, or local permitting programs. In other cases, the federal and state resource agencies agree on the use of a single permit application that circulates among all agencies, thus together developing a Joint Permitting Process.

For example, the USACE Jacksonville (FL) District and the Florida Department of Environmental Protection (FDEP) signed a State Programmatic General Permit (SPGP III-R1) to avoid duplication between both agencies for permitting minor work located in waters of the United States. This agreement covers shoreline stabilization (riprap, seawalls, and other shoreline stabilization structures, such as bulkheads). The applicant only submits an application to the FDEP office. In many cases, the FDEP processes the application without USACE input, i.e., "Green" cases. In these "Green" cases, no fill is placed more than 5 feet waterward of the MHW line; stabilization measures other than vertical seawalls are not steeper than 2H:1V [2 horizontal distances for 1 vertical distance]; and there are no adverse impacts on submerged aquatic vegetation or wetlands nor adverse impacts to federally listed endangered or threatened species. Cases not meeting these characteristics are classified as "Yellow" or "Red" and, therefore, involve different degrees of review by federal agencies.

The chapter has already mentioned other examples of intergovernmental coordination in programs that address shoreline erosion. Land use planning efforts by local governments often involve coordination with state planning agencies—either through delegation of authority or implementation of state land use guidelines. The national coastal zone management program is a partnership between the federal government that establishes program guidelines and the states that develop and implement their own coastal management programs with federal financial assistance and promises of "federal consistency." Other programs, such as EPA's National Estuary Program and NOAA's National Estuarine Research Reserve System are true federal-state partnerships with guidelines, significant financial support, and coordination originating at the federal level while implementation occurs at the state level.

Variability Among States

Although variations in the permitting processes occur between USACE districts, the greatest variation in permit requirements and constraints occurs among the states. These differences can be attributed to varying (1) geomorphologies and biological resources; (2) government organizational structures, i.e., USACE districts and regions are different from Environmental Protection Agency regions; state water quality and coastal zone certification authority may rest in the same or a different agencies; Native American interests may be significant; (3) historic and cultural differences; as well as (4) different institutional histories, cultures, and biases.

The following two examples, as well as Table 5-2, specifically illustrate the variations among states in their adoption of different shoreline protection options for sheltered coasts. Some states tend to favor armored responses while others create incentives for softer responses.

TABLE 5-2 Comparative Erosion Response for Sheltered Coasts by State

Methods to Address Erosion	Common	Occasionally	Rarely
Hardening (Bulkheads and Revetments)	SC, NC, FL, GA, AL, WA, VI, MS, LA, TX, MD (revetments), NJ, NY, PR	MA, MI, VA, DE	WA, MD (bulkheads)
Nourishment (Trap and Add Sand)	MA, VA, MD	FL, MI, TX, NJ, NY	AL, SC, WA, MI, MS, GA, NC, VI, LA, DE, PR
Alternatives Relying Predominantly on Vegetation	MA, LA, VA, MD	NC, WA, MI, FL, TX, DE	AL, SC, MS, GA, VI, NJ, NY, PR
Setbacks and Other Land Use Planning Techniques	MA, WA, MI, VA, DE, MD, NJ, NY	SC, NC, FL	GA, AL, VI, TX, LA, PR

NOTE: PR = Puerto Rico, VI = U.S. Virgin Islands.

North Carolina Example

North Carolina lies within the jurisdiction of four USACE districts and EPA Region 4. The North Carolina Department of Environment and Natural Resources (NC DENR) is the lead permitting agency for shoreline protection activities under a USACE regional permit. NC DENR houses both the state water quality and the coastal zone management programs. If the activity will occur in one of the twenty coastal counties within the jurisdiction of the Coastal Area Management Act (CAMA) and is along an estuarine or tidal shoreline, a CAMA permit is required. The NC DENR Division of Coastal Management has attempted to encourage alternatives to bulkheads, specifically sills, through adoption in 2005 of a General Permit for construction of riprap sills for wetlands enhancement in estuarine and public trust waters (15A NCAC 07H.2700). As with other general permits, it cannot be utilized if there are "unresolved questions concerning the proposed activity's impact on adjoining properties or on water quality, air quality, coastal wetlands; cultural or historic sites; wildlife; fisheries resources; or public trust rights" (www.nccoastalmanagement.net). This general permit requires consultation with the NC Division of Marine Fisheries and review of potential impacts on essential fish habitat. It is underutilized in part due to a shortage of professionals experienced in the design and installation of alternative erosion control structures.

Texas Example

Texas is located in EPA Region 6 and within the jurisdiction of four USACE districts. The Texas Coastal Management Program is housed in the state's General Lands Office and issues state permits for construction of shoreline protection structures, as well as making consistency determinations for USACE 404 permits. Most shoreline erosion projects are proposed to protect and create beaches impacted by shoreline erosion along sheltered shores. The state permits breakwaters, geotextile tubes, and beach nourishment. Texas also allows projects that create marsh habitat. However, bulkheads are easier to "get through the permitting process" and if time is a critical factor applicants are encouraged to move landward out of federal jurisdiction, thus bypassing the USACE permitting process. Additionally, activities that will occur in the water, such as breakwaters or beach nourishment, require a coastal lease or easement from the state prior to proceeding on state-owned lands.

To address the frustration of applicants with the complex and lengthy permitting process, a pilot program was initiated in Corpus Christi, TX. A Permitting Service Center was opened as a "one stop shop to help folks through the process." Staff at the Center assist applicants with the application process, assure all necessary information is included, and reduce permitting review time by limiting the need for regulatory review personnel to repeatedly contact applicants for additional information. The pilot project has been so successful that a second Permitting Service Center will open in Galveston. Some of the program's strongest proponents are USACE regulatory staff, who have commented that their review process has been shortened and improved because they are receiving the correct information with the original application.

Influence of the Regulatory Process

Not surprisingly, if property owners select a shoreline protection alternative that does not encroach into the highly regulated "waters of the United States," they can avoid significant transaction costs, lengthy permitting times, and numerous other aggravations. The regulatory implications of federal permits and the possibility of additional review under other legislation, sometimes create an incentive for the permit applicant to attempt to avoid the federal permit requirements by siting the project on uplands above the high tide line in coastal areas. State and local land use planning and coastal construction permits may still apply in these cases, but the applicant has simplified the regulatory panorama considerably by eliminating federal review. Of course, the suite of existing state and local regulatory requirements still apply, such as setback zones, stormwater management requirements, building codes, and vegetative buffers. This strong incentive to avoid or minimize encroachment into waters of the United States has created a bias both toward requesting and allowing certain erosion mitigation options, such as bulkheads and similar vertical structures. Thus, these structures may be constructed completely outside of the federally regulated area (i.e., above the MHW line and, therefore, not in navigable waters, as well as outside "waters of the United States") and still may meet a property owner's desired outcome of protecting upland properties. Constructing a bulkhead above the MHW line may be quicker and easier than obtaining a permit for a vegetative solution developed in the nearshore waters because it potentially avoids the multiple layers of federal review. In this way, the regulatory framework affects choices and outcomes.

The permitting process is generally reactive—rather than proactive. While the USACE has a long-established partnership with the states with respect to oceanfront responses to erosion, this partnership is much less developed for sheltered coasts. For example, the USACE's National Shoreline Erosion Control Development and Demonstration Program (Section 227 Program) is a research effort to construct, administer, and evaluate innovative and nontraditional coastal shoreline protection structures. The USACE and the states cooperate in site selection, administration, and evaluation. Most of the twelve Section 227 Program sites are located on the open-ocean coastline, rather than sheltered coasts (see further discussion of the Section 227 Program in Chapter 3). Comparisons exist between public policy and perceptions regarding erosion responses on the oceanfront versus on sheltered shorelines. During the 1980s, many communities and states focused on the appropriate response to *oceanfront* erosion. Significant long-term commitment to beach nourishment, as opposed to structures that hardened the shoreline, such as bulkheads, seawalls and revetments, was the norm. In some states, such as South Carolina, North Carolina, and Georgia, the permitting of these hard structures has now been prohibited on the oceanfront. Conversely, attention to the erosion issue on sheltered shorelines in the same states has yet to occur in any significant way, and hard structures are routinely permitted.

COST CONSIDERATIONS

The comparative costs of the erosion mitigation alternatives are an important factor in the decision-making process. However, these considerations include more than simply comparative construction costs. Some innovative alternatives may have high associated transaction costs that include additional time for approval, as well as additional studies and permit requirements.

The impacts of the technology may be beneficial or adverse on adjacent upland properties and nearshore lands, waters, and associated living resources. Although a challenging exercise, internalization of these potential benefits and adverse impacts into the decision-making process can help create a more realistic and coherent evaluation of costs and benefits.

Transaction and Administrative Costs of Shoreline Management

Transaction costs borne by property owners may include some or all of the following: site analysis, design, permitting, bidding, construction oversight, construction and monitoring of the chosen protection measure, as well as "lost" time. Protection measures may include no action, moving an endangered structure, nonstructural approaches (beach nourishment, zoning, vegetative), revetments, offshore breakwaters, groins, and bulkheads.

These costs are likely to be lower in jurisdictions where (1) a plan exists, (2) an administrative structure is in place that allows local, state, and/or federal officials to respond to individual requests for protection measures in the context of an overall plan, and (3) the regulatory structure creates an incentive to select certain options (i.e., general permits for certain structures). Costs will be higher if the protection measure requires federal permits—both the fees charged by the consultant and the time to achieve the desired outcome. If a project can be carried out solely under state and/or local permitting, the cost still varies with the size of the project and the protection measures sought.

Administrative costs occur at local, state, and/or federal levels. This category includes the permit technical review, permit completeness review, public hearings, meetings and consultations with project proponents, construction review, monitoring, potential litigation, and legislation.

Criteria to Assess the Costs of Shoreline Protection Measures

An evaluation of the economies of shoreline protection measures should be much broader than simply capital and operating costs. A more complete picture of criteria that should figure in the evaluation includes the following:

1. Capital costs
2. Operating costs, including monitoring and maintenance
3. Probability that the actions will reduce/eliminate erosion for X years
4. Impacts (positive and negative) on adjacent upland properties
5. Impacts (positive and negative) on public uses
6. Impact on environmental functions and thus on ecosystem services
7. Aesthetics

Items 1 through 5 can be assessed by competent professionals with experience in permitting erosion control techniques. Consideration of Item 6 has two parts: first, what must happen for these functions/services to be considered; second, how the evaluation is accomplished. It is likely that ecosystem values would be incorporated into the decision-making process when a proposed action spurs local groups (advocates for fisheries, water quality, bird or marine mammal habitat, for example) to appear at public hearings. The method for assigning values is described in *Valuing Ecosystem Services* (NRC, 2005), especially Chapter 4. The

NRC report points out that, while there is "growing recognition of the importance of ecosystem functions and services, they are often taken for granted and overlooked in environmental decision-making." The report further states that "the economic values of these ecosystems goods and services to society have to be known, so that they can be compared with the economic values of activities that may compromise them and improvements to one ecosystem can be compared to those in another" (NRC, 2005, p. 29).

Who Performs This Assessment and How?

All of the decision-makers mentioned earlier in this chapter have roles to play in the assessment of the various shore management options.

a. Property owners, working with their consulting engineers, must evaluate the cost (time and money) against the benefits (protection of desired features for some period of time).

b. Experts and consultants understand well the costs (capital, operating, administrative, and transaction) of various options, as well as the shoreline benefits. Many engineering consultants, however, may lack a clear understanding of the environmental costs and benefits of the various options.

c. Government regulators and compliance officials, working within a standard permitting process that allows for public comment, evaluate the proponents' desired choice of protection measure against local, state, and/or federal standards (either written or unwritten).

d. The USACE evaluates in the context of general permits and individual permits as described above.

e. Policy-makers do not usually engage in the evaluation of costs and benefits unless a particularly controversial permit has come to their attention.

It is inappropriate to require that an individual permit application include a cost-benefit analysis of alternative protection measures. This responsibility for evaluation of erosion control techniques across the same criteria should pertain to the regulatory agencies.

Potential Use of Mitigation Banks for Shoreline Erosion Control Permits

Wetland protection regulations began in the United States over 35 years ago when individual states and the federal government acted to protect the functions wetlands provide. While regulations vary from state to state, all allow wetland alteration under special conditions, and most jurisdictions require compensation or mitigation.

Banked wetlands are systems that have been restored or created in advance of the unavoidable impacts to wetlands that regulatory authorities do permit.

Ideally, banked wetlands should be managed, protected in perpetuity, functionally similar to the altered systems, and within defined geographical areas. Wetland mitigation banks are established under either or both state and federal guidance (Federal Register, 1995; FR 60(228) 58605). They have some or all of the following characteristics: a mission statement and agreement signed by regulators/other agency overseers, and bank sponsor; funding; a geographical scope (state, county, watershed, estuary, or bay); accounting procedures; methods for determining credits and debits; and policies on public review, land ownership, long-term land management; and wetlands types and conditions suitable for mitigation (USACE et al., 1998).

Compensation ratios may vary with type, size, and location of the wetland to be altered (Environmental Law Institute [ELI], 1994; Federal Register, 1995; FR 60(228) 58605; Zedler, 1996). As an example, one hectare (approximately two acres) of restored salt marsh credits may be required for each acre altered; two hectares (approximately five acres) of restored beach credits may be required for each acre altered. These ratios reflect local conditions and therefore will vary among jurisdictions.

At least 400 mitigation banks have been approved to provide credits, the majority of which have been established in the last decade to provide credits to the general public or permit applicant (Environmental Law Institute, 2006, USACE 2006b. This mitigation approach has caught the attention of the U.S. Congress, most recently in their call for rule-making for application of equivalent standards and criteria to each type of compensatory mitigation as part of the National Defense Authorization Act for Fiscal Year 2004 (Pub L., 108-136). This legislation, in effect, calls for other mitigation options, such as permittee-responsible mitigation and in-lieu fee mitigation (third party mitigation) to have standards equivalent to mitigation banking. EPA and USACE proposed revisions to their regulations on compensatory mitigation for losses of aquatic resources in March 2006 (71 FR 15519, 28 March 2006). Many consider wetland mitigation banking to be an evolving institutional mechanism (Environmental Law Institute, 1994; Marsh et al., 1996; Stein, 1999). For example, stream mitigation banking (for compensation for impacts to streams) and conservation banking (endangered species) approaches have been adopted in the last decade as part of the banking approach.

In some instances, it might be feasible to establish a mitigation bank for shoreline erosion projects. For example, some small individual projects that would adversely impact nearshore habitats could contribute to a mitigation fund or bank in compensation for the natural resources that they will degrade. The fund would accumulate contributions from many small projects and subsequently pay for a larger and more comprehensive restoration project than would have been possible with the small individual projects. The larger restoration project would ideally be located close to the affected area and would replicate the damaged resources—compensating for the damage to ecosystem services.

PLANNING CONSIDERATIONS

Few states have plans for their sheltered shorelines. Most regulatory agencies generally react to requests for shoreline alteration and seldom adopt strategies for determining desired shoreline features, habitat types, uses, and densities of uses. Generally, planning for responses to erosion occurs on a project-by-project basis on our sheltered shorelines. The major driver is the property owner's desire to protect his or her investment. Regulatory responses are usually associated with construction of new erosion control structures or the repair of existing structures.

Integrated planning efforts on sheltered coasts are increasing, however. Wetland restoration is very proactive and often involves quantitative performance goals. Shoreline protection is often one of the primary goals of estuarine habitat restoration.

For example, the loss of wetlands in Jamaica Bay (NY) is a visible shoreline erosion issue for the New York-New Jersey Harbor Estuary Program Office (HEP) of the USEPA. The HEP's research and pilot wetland restoration projects are not responses to permit applications for shoreline protection projects, but rather, stem from a desire to reverse the trend in wetland losses.

A number of additional National Estuary Program (NEP) sites have taken proactive roles addressing coastal erosion problems in areas under their authority. In Florida, the Tampa Bay Estuary Program has coordinated larger proactive shoreline erosion control efforts in association with its habitat restoration efforts. These efforts are not estuary-wide, however, but have occurred on stretches of shoreline several hundred meters in length.

The USACE possesses experience in conducting integrated planning in river basin, watersheds, and coastal systems—although the agency's use of integrated planning techniques could be improved (NRC, 2004b). The Corps' projects, such as the California Master Plan for Coastal Sediment Management, do not focus specifically on sheltered coastal systems nor on shoreline erosion. Nevertheless, the agency recognizes that "[s]ustainable solutions to beach erosion problems are not found through shoreline protection or beach nourishment on a local scale, but require a regional approach that links sediment sources and pathways to beach erosion and deposition."

Integrated planning approaches require an adequate information base. The applicant and/or planner must be cognizant of the specific nature of the erosion problem at the site and in the broader region (estuary, lagoon, or littoral cell). Current information and maps on estuarine erosion zones and rates are non-existent in many areas. For example, although the National Flood Insurance zone maps indicate high hazard flood zones, they do not consider shoreline erosion trends. As a result, property insurance premiums paid into this federal program do not reflect the risk of loss from shoreline erosion (Heinz Center, 2000). Besides information concerning high risk erosion sites, data concerning estuarine habitats and their ecological significance; shoreline density and development trends; and

potential sites for shoreline and habitat restoration can all be useful in addressing erosion.

The Surfrider Foundation [http://www.surfrider.org/stateofthebeach/home.
asp] annual reports show that states lack data on existing shoreline structures,
detailed habitat maps, or plans for shoreline protection. Some states still have
not mapped the erosion zones on their sheltered shores. It is often difficult for
the individual project applicant to obtain broad system-wide information. Basic
information for evaluating a site and selecting the most appropriate option will
include the following:

- Erosion history at the site and evidence of recent erosion activity
- Fetch
- General shape of the shoreline
- Shoreline orientation
- Slope and depth of the tidelands in front of the shoreline property
- Boat traffic
- Width of the beach above the MHT
- Bank height
- Bank composition
- Potential planting area
- Onshore gradient
- Beach vegetation below the project
- Shoreline and bank vegetation
- Existence of erosion control measures on adjacent properties

Several worksheets have been developed to assist in the evaluation of these
factors to determine Cumulative Erosion Potential Values (e.g., Riggs, 2001).

System planning also requires coordination among various regulatory agencies (federal, state, and local regulatory agencies), as well as government and
private landholders. A shared vision of the various social actors and public officials would form the basis for viable future shoreline plans.

System Planning

Integrated environmental planning involves several coordinated steps. The
initial phase requires an evaluation of the erosion problem at the site. It is necessary to determine the site conditions, the nature of the erosion problem, and its
causes. Greater amounts of appropriate information about the system will lead to
a more rational assessment of the needs, consideration of the best erosion mitigation options, and selection of the most appropriate alternative.

The "site" could be an individual parcel of land. However, advantages flow
from embracing a more holistic regional perspective rather than mere consideration of a single land holding. It might be better to conduct this type of planning

and analysis with a geographical scope that covers the entire estuary or embayment, the littoral or littoral cell, or an entire community. A coordinated community response could be the key to successful shoreline protection. Planning is performed at a scale that allows for coordinated response to a shoreline erosion phenomenon. A community response may also reduce the adverse impacts that one structure might have on adjacent properties.

A subsequent step for the property owner(s) or manager(s) would be to identify the management goals for the site and establish the priorities. These might include:

- Prevention of loss of taxable land
- Enhancement of water quality by managing runoff
- Protection of structures from loss to erosion
- Enhancement of coastal wetlands
- Provision of access to the beach
- Enhancement of recreational opportunities

The planner could then consider the most realistic options that would be most likely to meet the management goals. Alternative analysis involves a comparison of the different possible responses that the applicant is considering, the environmental costs and benefits of each, long-term durability of materials and lifetime of the project, the cost of the project, the likelihood of success of the efforts. Ease of permitting requirements might also be a consideration for the landholder or planner.

Cumulative Impact Analysis

The installation of any single shoreline erosion project may not significantly alter the local ecosystem, but the combination of many small projects over time may have a large impact on the types of habitats and ecosystem services available in the coastal zone. Under the National Environmental Policy Act (NEPA), these cumulative impacts are part of the evaluation of the range of policy alternatives included in environmental assessments and impact statements mandated for federal actions (Council on Environmental Quality, 1997). The Council on Environmental Quality defined cumulative effects in regulations for implementing NEPA as: "the impact on the environment which results from the incremental impact of the action when added to other past, present, and reasonably foreseeable future actions regardless of what agency (Federal or non-federal) or person undertakes such other actions (40CFR § 1508.7)." Assessment of cumulative effects is not just a federal requirement; many states also require cumulative impact analysis as part of environmental decision-making.

While system level planning can allow individual actions to be considered in a broader context, continual reassessment of the system state is necessary

---

BOX 5-3
Precautionary Approach

To ensure the sustainability of ecosystems for the benefit of future as well as current generations requires that decision makers follow a balanced precautionary approach, applying judicious and responsible management practices based on the best available science and on proactive, rather than reactive, policies. Where threats of serious or irreversible damage exist, lack of full scientific certainty should not be used as a justification for postponing action to prevent environmental degradation. Management plans and actions based on this precautionary approach should include scientific assessments, monitoring, mitigation measures to reduce environmental risk where needed, and periodic reviews of any restrictions and their scientific basis.

SOURCE: U.S. Commission on Ocean Policy, 2004.

---

because the cumulative impacts of actions within the system can have dramatic effects. For example, studies of the effects of development around lake shores on fish assemblages (Jennings et al., 1999) and habitat for northern pike and bluegill (Radomski and Goeman, 2001) have indicated the need to consider the cumulative effects of small habitat modifications. Although the effect of any individual protection measure on fish assemblages is difficult to identify during planning and permitting, or even post construction, the net effects of changes along the shoreline have been documented in many cases (e.g., Woodford and Meyer, 2003).

Cumulative effects may be an additive response to individual actions of the same type or the interactive consequence of multiple actions of different types (Spaling and Smit, 1993). Incorporating any potential cumulative effects of multiple actions into the planning process often requires a regulatory agency to look beyond issues within their own jurisdiction, as well as considering future, as yet unproposed, actions. Understanding the cumulative effects of shoreline protection measures within social, political, and ecological frameworks is an important component of an effective watershed or embayment plan. This provides a more holistic planning context that can reduce the unintended or unanticipated consequences of decision making on shoreline modification projects. In situations where insufficient information is available to provide an assessment of cumulative effects, a precautionary approach can be used to prevent irreversible loss of valuable habitats and other shoreline features (see Box 5-3).

FINDINGS

• There is an incentive to install seawalls, bulkheads, and revetments on sheltered coastlines because these erosion control structures can be built landward of the federal jurisdiction of the U.S. Army Corps of Engineers and thus avoid the need for federal permits.

• Existing biases of many decision-makers in favor of bulkheads and revetments with minimal footprints in public trust areas limit installation of other erosion control options that may provide more ecological benefits.

• The regulatory framework affects choices and outcomes. Producing a different outcome requires altering the incentives that emerge from the regulatory framework. Regulatory factors include the length of time required for the permit approval; incentives that the regulatory system creates to favor one technology over another; general knowledge of the options and understanding of the consequences, availability of information on the alternative technologies, planning support, and comparative costs.

• Traditional structural erosion control techniques may appear to be the most cost-effective. However, they do not account for the cumulative impacts that result in environmental costs nor the undervaluation of the environmental benefits of the nonstructural approaches.

• Nonstructural erosion control techniques provide both shoreline protection and ecosystem services (creation of fish habitat, habitat restoration, recreation benefits of nourished beaches).

• There is a general lack of knowledge and experience among decision-makers regarding options for shoreline erosion mitigation on sheltered coasts, especially options that retain more of the shoreline's natural features.

• The regulatory response to shoreline erosion on sheltered coasts is generally reactive rather than proactive. Most states have not developed plans for responding to and managing erosion on sheltered shorelines.

• Some states have not mapped the erosion zones on their sheltered shores, hindering informed decision-making by policy-makers and project applicants.

6

A New Management Approach for Sheltered Shorelines

Approaches for mitigating erosion on sheltered coasts have not been subject to the high level of scrutiny and national debate devoted to erosion on open coasts. Although erosion on sheltered coasts is not always analogous, state and local strategies for addressing erosion on ocean shorelines provide context for discussing new approaches to managing erosion on sheltered shorelines. In the 1980s to 1990s, appropriate options for managing erosion on ocean coasts formed the focus of national debate. Opinions ranged from "Properly engineered seawalls and revetments can protect the land behind them without causing adverse effects to the fronting beaches," (NRC, 1990) to "Hard stabilization may be the best way to save buildings, but retreating from the problem by removing buildings is the best way to save the beach" (Pilkey and Dixon, 1996). From these diverse views among coastal experts, improved approaches for managing ocean shorelines emerged based on studies of erosion processes and ocean front dynamics. One example, the regional sediment budget studies conducted by the U.S. Army Corps of Engineers (USACE, 2005), will be described in this chapter.

The current regulatory structure and prevailing erosion management approaches for sheltered shorelines have favored structures that harden the shoreline with the often unintended consequences of habitat loss, diminished resources, and recreational values. Implementation of a new management approach for sheltered shorelines could minimize further loss of the benefits associated with maintaining more natural shorelines. The prevailing practice of installing bulkheads and similar structures to combat real and perceived erosion problems is, for the most part, the inadvertent result of policies and regulations intended to protect upland property. Unfortunately, these policies sometimes do not give adequate consideration of the consequent loss of public trust resources such as beaches, wetlands, navigable waters, and submerged tidelands. Erosion mitigation measures have been developed that protect upland property and provide habitat and could be employed in a regional context to avoid undesirable cumulative impacts. A regional approach would help to account for the scale of erosion processes and facilitate exchange of information, technology, and experience gathered at the local level. More proactive planning for shoreline management will require the participation of local decision-makers in addition to state and federal agency partners. This chapter presents the components of a new regional framework for managing shore erosion, with recommendations based on the findings from this study.

REGIONAL APPROACHES

The term "regional" is used in this report to reflect an area of shoreline that is defined by some functional physical or ecological parameters such as littoral cells. Several examples of regional proactive planning already exist for shorelines: the USACE Regional Sediment Management approach, the USEPA National Estuary Program, and some special area management plans approved by state coastal management programs. Some aspects of all these experiences offer guidance for proactive regional planning for shoreline erosion control.

The U.S Army Corps of Engineers Regional Sediment Management (RSM) approach provides a model and framework that could be adapted to address sheltered shoreline erosion problems within a regional context. The RSM approach originated as a method for optimizing both economic and ecological resources. For decades, state and local managers of beach shorelines have been at odds with the U.S Army Corps of Engineers over navigation projects and the management of sediment. A demonstration program in 1999 in the northern Gulf of Mexico began a shift towards collaboration among federal, state, and local officials (Box 6-1). The RSM approach of the U.S Army Corps of Engineers has been essential to the agency's consideration of shoreline management at the regional scale. There are many factors in addition to sediment budgets to consider in the development of regional shoreline management plans. These factors include socioeconomic considerations as well as a broad range of habitat and other ecological issues. Regional plans facilitate the assessment of cumulative impacts and could be informed by credible monitoring of project performance and experience within and without the region of interest.

Developing a Regional Plan

Creating a Shoreline Vision

Environmental planning involves several coordinated steps. The initial phase requires coordination and involvement of all affected decision-makers at the

---

BOX 6-1
U.S. Army Corps of Engineers
Regional Sediment Management Approach

This approach enables the Corps, in partnership with state and local entities, to treat sediment as a resource in the context of a region (or littoral cell) through a systems approach to sediment management. According to the U.S Army Corps of Engineers, managing sediment to benefit a region potentially saves money, allows use of natural processes to solve engineering problems, and improves the environment. As a management method, RSM:

- Includes the entire environment, from the watershed to the sea
- Accounts for the effect of human activities on sediment erosion as well as its transport in streams, lakes, bays, and oceans
- Protects and enhances the nation's natural resources while balancing national security and economic needs. (USACE, 2005)

Sediment management encompasses erosion control measures, sediment removal through activities such as channel dredging, site and method of sediment deposition, and transport of material. Historically, these activities were conducted on a project-by-project basis in isolation of broader system considerations. Sediment management practices have ecosystem effects that extend beyond the local or individual space and time scale, and these cannot be considered unless the longer term and broader scale impacts of sediment management are considered. The RSM approach defines areas where sediment management actions will have a cumulative impact within a time of interest and with regard to the planned projects or actions. The RSM approach acknowledges the consequences of engineering projects and their impact on natural processes by considering longer time periods and larger areas than the immediate problem at a specific site. The success of the RSM approach depends on agency, intergovernmental, and stakeholder coordination and cooperation spanning political and geographical boundaries.

---

local, state, and federal levels. Although there are many factors that influence the state of a shoreline ecosystem, human activities are the most readily controlled and most easily planned. The degree to which human activities are controlled depends upon the development of a shared vision for the shoreline. This shared vision should include a long-term perspective consistent with the lifespan of the structures and activities to be governed by the plan. This vision is the first and most essential step in creating a regional plan and includes consideration of the economic, esthetic, and ecological values provided. Development of a shared vision will also incorporate the views of local property owners and users of the shoreline. In the absence of a vision to describe the desired shoreline, the future of sheltered shorelines will be decided through individual permitting decisions that may not reflect the values of the affected communities.

Establishing Goals

Establishing specific goals that are consistent with the shoreline vision is important to help guide decision-making and monitoring requirements, and provide a means to measure progress over time toward achieving the vision. The following are just some examples of shoreline management goals:

- Prevent loss of taxable land. Protect landward improvements and provide for personal safety.
- Enhance water quality by managing upland runoff and groundwater by maintaining wetland habitat and/or bioengineering of the bluff.
- Protect, maintain, enhance, or create wetlands and other intertidal habitat.
- Provide public access and create recreational opportunities by maintaining or creating beaches.
- Provide a sustainable coast that maintains existing uses in the face of rising sea level.
- Address potential or existing ecological impacts within the management area.
- Align costs with expected benefits to ensure the costs of shoreline management options are carefully analyzed in relation to their expected benefits.

The goals that are agreed upon apply within the context of a shoreline reach or littoral cell. Otherwise, not all mechanisms that could be contributing to the erosion problem will be addressed. Like creating the vision, it is important for all affected parties to work together to identify goals so that differences and potential conflicts can be resolved. Even if all goals are taken into account, different priorities will be given to different goals. It may not be possible to satisfy everyone if some of the goals for a given reach are mutually exclusive (Byrne and Zabawa, 1984).

Defining the Region

There are many ways to define the scope of the region of shoreline management plans. Typically erosion processes do not stop at political boundaries and may include two or more localities. Therefore, the preferred option for the geographic scope of the plan is to define the region based shoreline reach (shoreline segment) or littoral cell basis. This accomplishes two objectives: (1) it provides a focus on the erosion problem as opposed to political boundaries, and (2) it improves the ability to assess cumulative impacts. In cases where the physical setting spans political jurisdictions it obviously requires cooperative planning across these boundaries.

Using Good Information

A well thought out regional plan is based on the best system-wide information available on the social, economic, physical, and ecological resources of the
area. High quality information also facilitates the selection of the most appropriate erosion control alternatives. Maps of the region that illustrate shoreline
trends and describe key features (e.g., currents, man-made structures, wetlands,
sediment sinks and sources) provides a visual presentation of information that
is usually easier for the general public to understand. The location of land and
marine resources is also needed for decision-makers when contemplating whether
to avoid, minimize impacts, or enhance important resources. Good quality maps
also provide a better context and overview of the system, indicating trends of
shoreline development over time. Historical wind and water level data with
statistical return frequencies are also critical in assessing the wave climate and
the level of protection.

Evaluating Suitable Options for Addressing Erosion

A regional shoreline management plan presents a thorough discussion,
including cost and effectiveness, of the techniques and technologies available for
addressing erosion (i.e., land use management, vegetate, harden, and trap or add
sand). A "no action" alternative is also included. This is an important alternative
to consider, especially when assessing potential benefits such as sand supply to
shallow ecosystems. Many state or region-specific documents already provide a
sound foundation of information, but to include the full array of proven, available
options would require compilation and expansion. These summary documents
would need to be routinely updated to reflect knowledge gained from monitoring
and research activities for revisions of regional plans.

Monitoring Requirements

Erosion mitigation measures typically affect aquatic resources, landforms,
property values, private or public infrastructure, regional sediment supplies,
critical habitats on private properties, public property, or on property held in
public trust. Therefore, some level of monitoring is required to measure and
evaluate the effects on uplands and resources. Effective monitoring occurs at both
the individual project level (preconstruction baseline and more detailed assessments after project implementation) and the regional level covered by the plan.
Individual monitoring is typically the responsibility of project proponents while
regional monitoring is the responsibility of the management plan authority.

In addition to being consistent with the goals of the management plan, proponents, regulators and interested parties need to agree on "criteria for success"
that apply to project-level and regional-level activities. The monitoring plan is
then based upon these goals and criteria and the results of the monitoring are

used to measure progress toward the goals and criteria. Examples of "criteria for success" are presented in Box 6-2. Guidelines for selecting success criteria are available from U.S. Geological Survey (USGS, 1999), USACE (2002a), and previous National Research Council studies (2003, 2004b, 2005, 2006). Generally, success criteria are measurable, consistent with the purpose and goals of the project, and achievable by the end of a reasonable monitoring period (2-10 years). For example, success criteria in compensatory wetland mitigation projects have included percent canopy cover, percent plant survival, plant vigor, percent of native species, period of inundation, stability of designed hydrologic features, wildlife usage, and plant heights (USACE, 2004).

BOX 6-2
Monitoring Plan

A monitoring plan should include:

- Monitoring Methodology—address pre- and post-plan implementation or post-construction, depending on whether the monitoring is site or regional in scale; describe vis-à-vis success criteria; describe sampling methods—number, size, location, analytical tools; mapping/GIS.
- Monitoring Schedule—take into consideration the growing season (for vegetation), tidal or hydrology cycle to assess performance at times and intervals (months or years) appropriate to local conditions and the design of the project.
- Photos—ground and/or aerial photos taken from the same place (good reference points) every year to allow for interannual comparison.
- Reporting requirements and the ramifications of noncompliance.
- Adaptive Management—monitoring information is incorporated into ongoing regional or site-level management. There should be procedures in place to modify the project design in the event that the project does not meet the success criteria. Potential problems include loss of physical structures from storms, invasive vegetation, hydrological conditions (too wet/too dry), etc.
- Designation of responsible party and transferability of responsibility.
- Identification of who will do the monitoring, with a budget and a means to allocate the funds over the monitoring period.
- Identification of a site where the data will be archived and made available for analysis (i.e., a data repository).

SOURCE: Adapted from USACE (2004).

FINDINGS AND RECOMMENDATIONS

Establishing a New Management Approach

The following sections describe goals and provide recommendations for developing management approaches consistent with the unique ecological and physical processes of sheltered coasts.

GOAL: Maintaining natural features on sheltered coasts.

FINDING: The current regulatory framework for sheltered coasts contains disincentives to the development and implementation of erosion control measures that preserve more of the natural features of shorelines, mainly as a result of the combined lack of knowledge, vision, and planning. The existing system presents two main obstacles:

• Obstacle 1: General lack of knowledge and experience among decision-makers regarding alternative options for shoreline erosion response, the relative level of erosion mitigation afforded by the alternative approaches and their expected life time, and the nature of the associated impacts and benefits. This unfamiliarity with alternative engineering approaches has resulted in disinterest, concern, or disagreement among regulators regarding the ecological consequences of alternative shoreline stabilization measures.
• Obstacle 2: The current legal and regulatory framework discriminates against innovative solutions because of the complex and lengthy permitting process that almost always considers these options on a case-by-case basis.

Overcoming these obstacles would require a change in the current shoreline management framework. Decision-makers who are responsible for managing sheltered shorelines will need to become more educated about the potential solutions to shoreline erosion problems and take proactive steps to encourage approaches that minimize habitat loss and, if possible, enhance natural habitats. In addition, decision-makers will need to evaluate potential cumulative impacts of mitigation measures on shoreline features, habitats, or other amenities, but individual permit review, as currently practiced in most locales, does not include consideration of cumulative effects.

Virginia is implementing one example of a new proactive approach to managing sheltered coasts in the Chesapeake Bay. Virginia has established a permitting process that encourages alternatives to the use of vertical structures to stabilize shorelines if they are properly designed and installed.

This management change came in the form of a revision to an existing Virginia state statute to achieve no net loss of tidal wetlands. In May 2005, the Virginia Marine Resource Commission (VMRC) adopted a revised policy that, in part, removed the previous threshold of 1,000 square feet (approx. 90 square meters) for noncommercial projects (revised Wetland Mitigation-Compensation Policy and Supplemental Guidelines; Regulation 4 VAC 20-390-10 *et seq.*). These revisions address compensation for impacts of proposed bulkheads and revetments to vegetated and nonvegetated wetlands. As revised, any proposed impact to tidal wetlands will be addressed by compensation through mitigation, wetland banking, or use of in-lieu fees, in that order. The use of properly engineered and installed marshes and beaches for shore stabilization may be deemed acceptable onsite compensation or mitigation by VMRC.

FINDING: Development of an integrated plan for management of shore erosion will require improvements in the scope and accessibility of the available information, including the nature of the erosion problem at specific sites and the overall patterns of erosion, accretion, and inundation in the broader region (estuary, lagoon, littoral cell). Many areas lack geospatial information on estuarine erosion zones and rates despite the relevance to coastal planning and its importance in the assessment of vulnerability to natural hazards. For example, although the National Flood Insurance zone maps indicate high hazard flood zones, they do not consider shoreline erosion trends. As a result, property insurance premiums paid into this federal program do not reflect the risk of loss from shoreline erosion, thereby underestimating the actual risk of flooding. This, in effect, subsidizes and encourages the development (or rebuilding) of structures in high-risk areas.

RECOMMENDATIONS:
   • **State and federal agencies (EPA, USACE, and NOAA) need to convene a working group to evaluate the decision-making process used for issuing permits for erosion mitigation structures to revise the criteria for sheltered coasts, including consideration of potential cumulative impacts.**
   • **Proactive erosion mitigation plans should be implemented to avoid the unintended consequences from hardened shorelines that reduce the recreational, esthetic, economic, and ecological value of sheltered coastal areas. The regulatory preference for permitting bulkheads and similar structures could be changed to favor more ecologically beneficial solutions that still mitigate erosion.**
   • **Long-term shoreline erosion information should be gathered and included in publicly available maps, such as flood insurance rate maps, to more accurately reflect the risks and potential costs of building on erosion-prone shorelines. These maps would provide better information for property buyers and insurers and could guide localities in establishing land use and zoning policies.**
   • **The USGS should include data collection and reporting on sheltered shorelines as part of their National Assessment of Coastal Change program**

GOAL: Understanding sheltered shoreline processes and ecological services.

FINDING: Decision-makers need adequate information on the effectiveness of new techniques for addressing erosion and the effects of various mitigation strategies on the physical and biological characteristics of the affected areas to achieve a more balanced approach to erosion along sheltered coasts.

Although there are a few good examples of publications that address the physical process of sheltered coast systems (Finlayson and Shipman, 2003; Jackson et al., 2002; Riggs, 2001), overall, less is known about these systems than about open coasts. Basic information, such as resource characterization, shoreline change analysis, sediment transport patterns, habitat function, and ecological services, is available for only a portion of the nation's sheltered shorelines and few programs address these knowledge gaps. States have not committed the resources necessary to periodically collect and analyze data for a comprehensive assessment of affected shorelines as would be necessary for effective regional planning.

Also, decision-makers (especially property owners) need assessments of new techniques and materials designed to mitigate shore erosion. Because of the comparatively low energy environments on sheltered coasts, special techniques have been developed to address erosion in these areas. Some techniques, such as the combination of a planted marsh fringe with a sill have been tested and proven effective under well characterized physical settings. However, new techniques (or structural materials) are periodically introduced that require a rigorous process of testing and evaluation to determine their effectiveness in controlling erosion and to evaluate their impacts on the environment.

RECOMMENDATIONS:
• Federal agencies (e.g., USACE, EPA, USGS, NOAA, and NSF), state agencies, and coastal counties and communities should support targeted studies of sheltered coast dynamics to provide an informed basis for selecting erosion mitigation options that consider the characteristics of the broader coastal system rather than simply addressing immediate problems at individual sites. Topics for studies:

— Identify trade-offs in ecosystem services associated with various mitigation measures,
— Quantify the costs and benefits of nonstructural erosion control techniques,
— Document system-wide process and hazard information, including mapping of erosion zones and rates. This information needs to be presented in nontechnical formats such as summary maps that can be readily understood by decision-makers, and
— Develop models to predict the evolution of coastal features under various scenarios.

- State and federal agencies should ensure that the information obtained from these studies is readily available to decision-makers at all levels of government.
- State and federal regulatory programs should establish a technical assistance function to provide advice on permitting issues and information on types of erosion mitigation approaches and their effectiveness under various site conditions.

GOAL: Improving awareness of alternative measures for addressing erosion.

FINDING: Many decision-makers, particularly homeowners but also some state and federal regulators, are not sufficiently informed about the choices available to them or the short and long term impacts of their choices. Chapter 3 presents many different approaches and technologies that address erosion along sheltered coasts. Inconsistency among states, and even USACE districts, in terms of permitting and the availability of information about various mitigation measures, unnecessarily constrains options and outcomes.

RECOMMENDATIONS:
- The major federal agencies involved in permitting activities (EPA, USACE, and NOAA) should initiate a national policy dialogue on erosion mitigation for sheltered coasts to bring together state and federal decision-makers and share information on the potential application and value of different mitigation approaches.
- The national dialogue should be used to develop guidelines for mitigating erosion on sheltered coasts that give deference to ecologically beneficial measures and ensure consistency of decision-making across regions.
- The national dialogue will require development of handbooks or Web pages with objective information about erosion mitigation techniques, including descriptions of the conditions under which each option would be effective. These handbooks (or Web pages) should be actively distributed to state and local planning and permitting staff; professional associations of environmental consultants, engineers, zoning officials, planners, and building inspectors; and extension agents; and made readily available to property owners and community groups.
- Professional societies and conferences should be utilized as a venue for transferring information to decision-makers such as regulators, engineers, and consultants.

GOAL: Document individual and cumulative effects of erosion mitigation approaches.

FINDING: Cumulative effects of shoreline hardening are rarely assessed and hence are generally unknown. However, an appreciation of the potential cumulative effects will be necessary to prevent an underestimation of the impacts of individual projects. As discussed in Chapters 4 and 5, all shoreline management options have associated costs and benefits. To effectively evaluate these costs and benefits requires delineation of the area of influence of the proposed option and assessment of the associated cumulative impacts. Generally speaking, the physical area influenced by mitigation structures is limited to adjacent properties and their adjoining ecosystems in the case of small structures like bulkheads or to the littoral cell in the case of large structures such as groin fields. The societal area of influence can be much more extensive because armoring projects tend to initiate community trends and preferences for these types of structures along shorelines.

Cumulative impacts encompass legal, social, and physical effects. From a legal or regulatory perspective, issuance of a permit may establish a precedent, potentially facilitating the approval process for future requests for similarly situated structures. This is one form of cumulative impact that often results in similar shoreline structures being constructed throughout a regulatory jurisdiction. Another aspect of cumulative impact is the erosion enhancing effect of structures such as bulkheads on adjoining properties. Flanking property owners are likely to respond by constructing their own bulkheads, with a domino-type effect up and down the shoreline.

Although loss of small parcels of shoreline habitat from hardening may not have a large impact on the ecosystem, the cumulative impact of the loss of many small parcels will at some point alter the properties and composition of the ecosystem. For example, sand- or mud-dependent intertidal species could be replaced by species that favor hard substrates such as rocky revetments. In addition, the economic, recreational and esthetic properties of the shoreline will be altered, with potential loss of public use, access, and scenic values. However, it is difficult to identify the point at which individual projects accumulate to an extent that threatens the valued properties of the shoreline. This requires a determination of the values the affected communities invest in the nonhardened shoreline and an assessment of the value of the ecosystem properties that stand to be lost. Both require a significant amount of information that might not be immediately available to decision-makers. As noted in Chapter 5, regulators are responsible for protecting the public trust values of the ecosystems that are affected by their decisions, and this requires an evaluation of the impact of permit decisions on ecosystem services.

RECOMMENDATIONS:
   • **The decision-making process should account for ecosystem services when permitting coastal shoreline stabilization projects. Decision-makers**

require objective, quantifiable methods for evaluating specific ecosystem services as one of the criteria in permit decisions.

• The economic, recreational, and esthetic properties of the shoreline should be evaluated to assess potential loss of public use, access, and scenic values.

• Cumulative effects should be considered in shoreline management plans, both for the values invested by the affected communities in non-hardened shorelines and the value of ecosystem properties that stand to be lost with shoreline hardening. Although it may not be possible to identify the threshold beyond which cumulative impacts become unacceptable or irreversible, anticipation of the problem allows prioritization of projects in areas unsuited to nonstructural alternatives or sites where structures are predicted to have less impact.

• In the absence of a comprehensive assessment of the cumulative impacts of erosion mitigation measures, a precautionary approach should be used to prevent the unintentional loss of shoreline features and significant alteration of the coastal ecosystem.

GOAL: Proactive shoreline management planning.

FINDING: The current permitting system fosters a reactive response to the problem of erosion on sheltered coasts (see Chapter 5). Decision-making is usually parcel-by-parcel and based on little or no physical or ecological information. The path of least resistance drives choices through a rigid decision-making process, with inadequate attention to the cumulative effects of individual decisions.

Creating a more proactive "regional approach" to shoreline management could address the unintended consequences of reactive permit decisions.

RECOMMENDATIONS:
• The development of regional shoreline management plans for sheltered coasts should occur at the state and local level in partnership with the federal government. Plans should be proactive and comprehensive in scope, and should be scaled to the estuary, bay, or littoral cell as appropriate.

• Regional shoreline management plans should include the following essential elements: (1) a shared vision and goals for the future shoreline of the water body through stakeholder collaboration, (2) analysis of regional sediment budgets and the cumulative effects of existing shoreline management activities, (3) the mechanism for turning the vision into reality through consistent permitting provisions, (4) implementation, and (5) performance evaluation and monitoring requirements (adaptive management).

• Plans should be considered "living documents" and updated every 5 to 10 years as new information (e.g., monitoring data, research results) becomes available.

- **Each regional shoreline management plan should describe the physical and hydrodynamic settings, including the location and type of existing shoreline structures in a GIS format. The plan should describe the available mitigation options and discuss the applicability, relative cost and benefit, and effectiveness of each option.**
- **Monitoring should include both a preconstruction baseline and more detailed assessments after project implementation, both at the individual project level and for the entire region covered by the plan. Individual monitoring should be the responsibility of project proponents while regional monitoring should be the responsibility of the management plan authority.**
- **Information obtained from monitoring programs should be incorporated in subsequent planning activities to support adaptive management as a mechanism to consistently evaluate and refine regional plans.**

Regional planning generally takes place at the state and local level in partnership with the federal government. To be effective, these planning efforts should involve property owners and other stakeholders early in the process. The programs most suited to undertake planning for regional shoreline management are the state coastal zone management programs. Shoreline management plan development could be an eligible activity under Section 309 of the Coastal Zone Management Act and could be effectively implemented as Special Area Management Plans (SAMPs). Creating regional shoreline management plans under the auspices of SAMPs would also provide an opportunity to employ the federal consistency provisions of the Coastal Zone Management Act (CZMA) to ensure that federal permitting actions are consistent with the plan (see Box 6-3).

Applying principles of adaptive management to regional shoreline management plans will allow these plans to be continually updated and improved as new information becomes available. Adaptive management acknowledges the uncertainty of systems; therefore, it integrates monitoring and evaluation into an iterative decision-making process for management. Essential elements of an adaptive management plan can be found in *Adaptive Management for Water Resources Project Planning* (NRC, 2004a).

CONCLUSION

Overcoming the obstacles associated with the existing management framework will require a number of societal and institutional changes: (1) better understanding of sheltered shoreline processes and ecological services; (2) improved awareness of the choices available for erosion mitigation; (3) documentation of individual and cumulative consequences of erosion mitigation approaches; (4) proactive shoreline management planning that takes into consideration the

---

BOX 6-3
Coastal Zone Management

The National Coastal Zone Management (CZM) Program is a voluntary part-nership between the federal government and U.S. coastal states and territories authorized by the Coastal Zone Management Act of 1972 (16 U.S.C. §§1451-1466) to:

- Preserve, protect, develop, and, where possible, restore and enhance the resources of the nation's coastal zone for this and succeeding generations;
- Encourage and assist the states to exercise effectively their responsibilities in the coastal zone to achieve wise use of land and water resources there, giving full consideration to ecological, cultural, historic, and esthetic values, as well as the need for compatible economic development;
- Encourage the preparation of special area management plans to provide increased specificity in protecting significant natural resources, reasonable coastal-dependent economic growth, improved protection of life and property in hazardous areas and improved predictability in governmental decision-making; and
- Encourage the participation, cooperation, and coordination of the public, federal, state, local, interstate and regional agencies, and governments affecting the coastal zone.

Since 1974, with the approval of the first state CZM program in Washington (Washington State Department of Ecology, 2006), a total of 34 coastal states and five island territories have developed CZM programs. Together, these programs protect more than 99 percent of the nation's 153,400 kilometers (approx. 95,331 miles) of oceanic and Great Lakes coastline.

The federal consistency provision with approved state coastal management plans is an important incentive for states to develop their plans. The federal govern-ment promises that federal agency activities, federally permitted or licensed activi-ties, or outer continental shelf exploration, development, or production activities will be consistent with the enforceable policies of the approved state management programs.

SOURCE: NOAA, 2005.

---

unique ecological and physical processes of sheltered coasts; and (5) a permitting system with incentives that support the goals of the shoreline management plan. The objective is an erosion mitigation decision-making process that helps achieve the shoreline management plan.

References and Bibliography

Airoldi, L., M. Abbiati, M.W. Beck, S.J. Hawkins, P.R. Jonsson, D. Martin, P.S. Moschella, A. Sundelof, R.C. Thompson and P. Aberg. 2005. An ecological perspective on the development and design of low-crested and other hard coastal defense structures. *Coastal Engineering* 52:1073-1087.

Allen, H.H., R.L. Lazor and J.W. Webb. 1990. Stabilization and development of marsh lands. Beneficial Uses of Dredged Material. Pp. 101-112 in *Proceedings of the Gulf Coast Regional Workshop*, held on April 26-28, 1988 in Galveston, Texas. Technical Report D-90-3. U.S. Army Corps of Engineers, Waterways Experiment Station, Vickburg, Mississippi.

Amos, C.L., M. Brylinsky, T.F. Sutherland, D. O'Brien, S. Lee and A. Cramp. 1998. The stability of a mudflat in the Humber estuary, South Yorkshire, UK. Pp. 25-44 in *Sedimentary processes in the intertidal zone*, Black, K.S., D.M. Paterson, and A. Cramp (eds). Geological Society of London Special Publication 139, United Kingdom.

Badola, R. and Hussain, S.A. 2005. Valuing ecosystem functions: an empirical study on the storm protection function of Bhitarkanika mangrove ecosystem, India. *Environmental Conservation* 32(1):85-92.

Bailard, J.A. and D.L. Inman. 1981. An energetics bedload transport model for a plane sloping beach: Local transport. *Journal of Geophysical Research* 86:2035.

Bak, P. 1996. *How Nature Works: The Science of Self-organized Criticality*. Copernicus, New York, New York.

Basco, D.R. and C.S. Shin. 1993. *Design Wave Information for Chesapeake Bay and Major Tributaries in Virginia*. Report no 93-1. Old Dominion University, Costal Engineering Institute, Norfolk, Virginia.

Beatley, T., D.J. Brower and A.K. Schwab. 2002. *An Introduction to Coastal Zone Management*. Island Press: Washington, DC

Beck, M.W., K.L. Heck, Jr., K.W. Able, D.L. Childers, D.B. Eggleston, B.M. Gillanders, B.S. Halpern, C.G. Hays, K. Hoshino, T.J. Minello, R.J. Orth, P.F. Sheridan and M.P. Weinstein. 2003. The Role of Nearshore Ecosystems as Fish and Shellfish Nurseries. Issues in Ecology 11:2-12. [Online] Available at: http://www.esa.org/science/Issues/ [January 31, 2006].

Bedford, B.L. 1996. The need to define hydrologic equivalence at the landscape scale for freshwater wetland mitigation. *Ecological Applications* 6:57-68.

Borges, P., C. Andrade and M.C. Freitas. 2002. Dune, bluff and beach erosion due to exhaustive sand mining: the case of Santa Barbara Beach, São Miguel (Azores, Portugal). *Journal of Coastal Research*, Special Issue 36:89-95.

Bowen, A.J. 1969. The generation of longshore currents on a plane beach. *Journal of Marine Research* 27:206-215.

Bowen, A.J. 1980. Simple models of nearshore sedimentation; beach profiles and longshore bars. Pp. 1-11 in *The Coastline of Canada*, S.B. McCann (ed.). Geological Survey of Canada, Ottawa, Ontario, Canada.

Brinson, M.M. and R. Rheinhardt. 1996. The role of reference wetlands in functional assessment and mitigation. *Ecological Applications* 6:69-76.

British Permanent Service. 2005. *Permanent Service for Mean Sea Level*. [Online]. Available at: http://www.pol.ac.uk/psmsl/ [November 29, 2005].

British Standards Institution. 1991. *British Standard for Maritime Structures, Part 7: Guide to the design and construction of breakwaters*. British Standards Institution, London, United Kingdom.

Bruun, P. 1962. Sea-level rise as a cause of shore erosion. *Journal Waterways and Harbours Division* 88(1-3):117-130.

Burby, R.J. (ed.). 1998. *Cooperation with Nature: Confronting Natural Hazards with Land-Use Planning for Sustainable Communities*. John Henry Press/National Academy Press: Washington, DC

Burby, R.J. and R.G. Paterson. 1993. Improving compliance with state development regulations. *Journal of Policy Analysis and Management* 12:753-772.

Burby, R.J., R.E. Doyle, D.R. Godschalk, and R.B. Olshansky. 2000. Creating hazard resistant communities through land-use planning. *Natural Hazards Review* 1:99-106.

Byrne, R.J. and C. Zabawa. 1984. *Effects of Boat Wakes on Tidal Shorelines*. Technical Report. College of William and Mary, Virginia Institute of Marine Science, Gloucester Point.

Carminati, E. and G. Martinelli. 2002. Subsidence rates in the Po Plain, northern Italy: The relative impact of natural and anthropogenic causation. *Engineering Geology* 66:241-255.

Clark, W. 2001. *Protecting the Estuarine Region through Policy and Management*. North Carolina Sea Grant, Raleigh.

Costanza, R., R. d'Arge, R. de Groot, S. Farber, M. Grasso, B. Hannon, K. Limburg, S. Naeem, R.V. O'Neill, J. Paruelo and R.G. Raskin. 1997. The value of the world's ecosystem services and natural capital. *Nature* 387:253-260.

Coulten, K., D. Divoky, D. Hatheway, J. Johnson and R. Nobel. 2005. *Sheltered Waters: FEMA Coastal Flood Hazard Analysis and Mapping Guidelines Focused Study Report*. Federal Emergency Management Agency, Washington, DC.

Council on Environmental Quality. 1997. *Considering Cumulative Effects under the National Environemtal Policy Act*. Executive Office of the President, Council on Environmental Quality, Washington, DC.

Craft, C., J. Reader, J.N. Sacco and S.W. Broome. 1999. Twenty-five years of ecosystem development of constructed *Spartina alterniflora* (Loisel) marshes. *Ecological Applications* 9:1405-1419.

Dally, W.R and J. Pope. 1986. *Detached breakwaters for shore protection*. Technical Report CERC-86-1. U.S. Army Corps of Engineers, Vicksburg, Mississippi.

Dan, A., A. Moriguchi, K. Mitsuhashi and T. Terawaki. 1998. Relationship between *Zostera marina* and bottom sediments, wave action offshore, in Naruto, Southern Japan. *Fisheries Engineering* 34:229-204.

Danielsen, F., M.K. Sorensen, M.F. Olwig, V. Selvam, F. Parish, N.D. Burgess, T. Hiraishi, V.M. Karunagaran, M.S. Rasmussen, L.B. Hansen, A. Quarto and N. Suryadiputra. 2005. The Asian tsunami: a protective role for coastal vegetation. *Science* 310:643.

Dennison, W.C., R.J. Orth, K.A. Moore, J.C. Stevenson, V. Carter, S. Kollar, P. Bergstrom and R. Batiuk. 1993. Assessing water quality with submersed aquatic vegetation. Habitat requirements as barometers of Chesapeake Bay health. *Bioscience* 43:86-94.

Douglass, S.L. 2005a. *How bad is the Problem: Case Study of Mobile Bay.* Presentation to the Committee on Mitigating Shore Erosion along Sheltered Coasts, meeting held on October 4-6, 2005 in Seattle, Washington. National Research Council, Washington, DC.

Douglass, S.L. 2005b. *The Tide Don't Go Out Anymore.* Presentation to the Committee on Mitigating Shore Erosion along Sheltered Coasts, meeting held on October 4-6, 2005 in Seattle, Washington. National Research Council, Washington, DC.

Douglass, S.L. and B.H. Pickel. 1999. *"The Tide Doesn't Go Out Anymore"—The Effect of Bulkheads on Urban Bay Shorelines.* University of South Alabama, Civil Engineering and Marine Sciences Departments, Mobile. [Online] Available at: http://www.southalabama.edu/cesrp/Tide.htm [January 25, 2006].

Dring, M.J. 1992. *The Biology of Marine Plants.* Cambridge University Press, Cambridge, United Kingdom. 199 pp.

Duke University. 2005. *Description of Alternative Devices for Shoreline Stabilization.* [Online] Available at: http://www.nicholas.duke.edu/psds/Stabilization/Categories.htm [November 29, 2005].

Ehrenhauss, S., U. Witte, F. Janssen and M. Huettel. 2004. Decomposition of diatoms and nutrient dynamics in permeable North Sea sediments. *Continental Shelf Research* 24: 721-737.

Environmental Law Institute (ELI). 1994. *National wetland mitigation banking study.* IWR Report 94-WMB-2. U.S. Army Corps of Engineers, Institute for Water Resources, Alexandria, Virginia.

Environmental Law Institute (ELI). 2006. 2005 *Status Report on Compensatory Mitigation in the United States.* An ELI Report prepared by Wilkinson and Thompson. Environmental Law Institute, Washington, DC.

Eurosion. 2004. *Living with coastal erosion in Europe: Sediment and space for sustainability. Part IV—A guide to coastal erosion management practices in Europe: Lessons learned.* Eurosion, Directorate General Environment European Commission, Hague, The Netherlands [Online] Available at: http://www.eurosion.org/reports-online/part4.pdf [September 7, 2006].

Field, C.D. 1997. *Journey Amongst Mangroves.* International Society for Mangrove Ecosystems, University of the Ryukus, Okinawa, Japan.

Finlayson, D.P. and H. Shipman. 2003. Puget Sound Drift Cells: the importance of waves and wave climate. *Puget Sound Notes: Science News from the Puget Sound Action Team* 47:1-4.

Fonseca, M.S. and J. A. Cahalan. 1992. A preliminary evaluation of wave attenuation for four species of seagrass. *Estuarine, Coastal and Shelf Science* 35:565-576.

Fonseca, M.S., W.J. Kenworthy and G.W. Thayer. 1998. *Guidelines for the conservation and restoration of seagrasses in the United States and adjacent waters.* National Oceanic and Atmospheric Administration (NOAA) Coastal Ocean Program Decision Analysis Series. No. 12. NOAA, Coastal Ocean Office, Silver Spring, Maryland. 222 pp.

Fonseca, M.S., P.E. Whitfield, N.M. Kelly and S.S. Bell. 2002. Modeling seagrass landscape pattern and associated ecological attributes. *Ecological Applications* 12:218-237.

Fredsøe, J. and R. Deigaard. 1992. *Mechanics of Coastal Sediment Transport.* World Scientific Publishing Company, River Edge, New Jersey. 369 pp.

Gabriel, A.O. and T.A. Terich. 2005. Cumulative patterns and controls of seawall construction, Thurston County, Washington. *Journal of Coastal Research* 21:430-440.

Great Lakes Basin Commission. 1976. Appendix 12: Shore Use and Erosion, in *Great Lakes Basin Framework Study,* prepared by Shore Use and Erosion Work Group and sponsored by U.S. Army Corps of Engineers, North Central Division. Great Lakes Basin Commission, Ann Arbor, Michigan.

Griggs, G.B. and K.B. Patsch. 2004. California's Coastal Cliffs and Bluffs. Pp. 53-64 in *Formation, Evolution, and Stability of Coastal Cliffs—Status and Trends*, Hampton, M.A. and G.B. Griggs (eds.). Professional Paper 1693. U.S. Department of the Interior, U.S. Geological Survey. United States Government Printing Office, Washington, DC.

Griggs, G.B. and L.E. Savoy. 1985. *Living with the California Coast*. Duke University Press, Durham, North Carolina. 393 pp.

Griggs, J.F. and G.B. Tait. 1991. *Beach Response to the Presence of a Seawall; Comparison of Field Observations*. Contract Report CERC 91-1. U.S. Army Corps of Engineers, Waterways Experiment Station, Vickburg, Mississippi.

Griggs, G.B., J.E. Pepper and M.E. Jordan. 1992. *California's coastal hazards; a critical assessment of existing land-use policies and practices*. Special Publication of California Policy Seminar Program, Berkeley, California. 224 pp.

Griggs, G.B., J.F. Tait and W. Corona. 1994. The Interaction of Seawalls and Beaches; Seven Years of Monitoring, Monterey Bay, Califorinia. *Shore and Beach* 62(3):21-28.

Habel, J.S. and G.A. Armstrong. 1978. Assessment and atlas of shoreline erosion along the California coast. State of California, Department of Navigation and Ocean Development. Sacramento.

Hardaway, C.S., Jr. and R.J. Byrne. 1999. *Shoreline Management in Chesapeake Bay*. Special Report in Applied Marine Science and Ocean Engineering, No. 356. College of William and Mary, Virginia Institute of Marine Science, Gloucester Point. 54 pp.

Hardaway, C.S., Jr. and R.J. Byrne. 2001. *Shoreline Management in Chesapeake Bay*. College of William and Mary, Virginia Institute of Marine Science, Gloucester Point.

Hardaway, C.S., Jr. and J.R. Gunn. 1999. Chesapeake Bay: Design and Early Performance of Three Headland Breakwater Systems. Pp. 828-843 in *Coastal Sediments '99, Proceedings of the 4th International Symposium on Coasting Engineering and Science of Coastal Sediment Processes*, held on June 21-23, 1999, in Hauppauge, New York. American Society of Civil Engineers, Reston, Virginia.

Hardaway, C.S., Jr., G.R. Thomas, M.A. Unger, J. Greaves and G. Rice. 1991. *SEABEE Monitoring Project, James River Estuary, Virginia*. Report for the Center for Innovative Technology by Virginia Institute of Marine Science. College of William and Mary, Virginia Institute of Marine Science, Gloucester Point.

Hardaway, C.S., Jr., J. Posenau, G.A. Thomas and J.C. Baumer. 1992. *Shoreline Erosion Assessment Software (SEASware) Report*. Technical report to the Virginia Department of Conservation and Recreation. Richmond, Virginia. College of William and Mary, Virginia Institute of Marine Science, Gloucester Point. 48 pp. + Appendices.

Hardaway, C. S., Jr., J.R. Gunn and R.N. Reynolds. 1993. Breakwater Design in the Chesapeake Bay: Dealing with the End Effects. Pp. 27-41 in *Coastal Engineering Considerations in Coastal Zone Management, Proceedings of the 8th Symposium on Coastal and Ocean Management*, held on July 19-23, 1993 in New Orleans, Louisiana. American Society of Civil Engineers, Reston, Virginia.

Hardaway, C.S., Jr., D.A. Milligan, C.A. Wilcox, L.M. Meneghini, G.R. Thomas and T.R. Comer. 2005. *The Chesapeake Bay Breakwater Database Project: Hurricane Isabel Impacts to Four Breakwater Systems*. College of William and Mary, Virginia Institute of Marine Science, Gloucester Point.

Heck, K.L. Jr., K.W. Able, C.T. Roman and M.P. Fahay. 1995. Composition, abundance, biomass, and production of macrofauna in a New England estuary: Comparisons among eelgrass meadows and other nursery habitats. *Estuaries* 18:379-389.

Heinz Center. 2000. *Evaluation of Erosion Hazards*. Heintz Center, Washington, DC. [Online] Available at: http://www.heinzctr.org/publications.htm [April 7, 2006].

Hesp, P. 2002. Foredunes and blowouts: initiation, geomorphology and dynamics. *Geomorphology* 48:245-268.

Hjulstrom, F. 1939. Transportation of detritus in moving water. Pp. 5-31 in *Recent marine sediments*, P.D. Trask (ed.). American Association of Petroleum Geologists, Tulsa, Oklahoma.

Holway, J.M. and R.J. Burby. 1993. Reducing flood losses: Local planning and land use controls. *Journal of the American Planning Association* 59:202-216.

Iannuzzi, T.J., M.P. Weinstein, K.G. Sellner and J.C. Barrett. 1996. Habitat disturbance and marina development: an assessment of ecological effects. 1. Changes in primary production due to dredging and marina construction. *Estuaries* 19:257-271.

Intergovernmental Panel on Climate Change (IPCC). 2001. *Climate Change 2001: The Scientific Basis*, Houghton, J.T., X. Dai, Y. Ding, D.J. Griggs, C.A. Johnson, P.J. van der Linden, K. Maskell and M. Noguer (eds.). Cambridge University Press, Cambridge.

Jackson, N.L. 1996. Stabilization on the Shoreline of Raritan Bay, New Jersey. Pp. 397-420 in *Estuarine Shores: Evolution, Environments and Human Alterations*, K.F. Nordstrom and C.T. Roman (eds.). John Wiley & Sons, Chichester, United Kingdom.

Jackson, N.L., K.F. Nordstrom, I. Eliot and G. Masselink. 2002. "Low Energy" Sandy Beach in Marine and Estuarine Environments: a review. *Geomorphology* 48:147-162.

Jennings, M.J., M.A. Bozek, G.R Hatzenbeler, E.E. Emmons and M.D. Staggs. 1999. Cumulative Effects of Incremental Shoreline Habitat Modification on Fish Assemblages in North Temperate Lakes. *North American Journal of Fisheries Management* 19:18-27.

Kalo, J.J., R.G. Hildreth, A. Rieser, D.R. Christie and J.L. Jacobson. 2002. *Coastal and Ocean Law*. West Group, St. Paul, Minnesota.

Kemp, W.M., R. Batuik, R. Bartleson, P. Bergstrom, V. Carter, G. Gallegos, W. Hunley, L. Karrh, E. Koch, J. Landwehr, K. Moore, L. Murray, M. Naylor, N. Rybicki, J.C. Stevenson and D. Wilcox. 2004. Habitat requirements for submerged aquatic vegetation in Chesapeake Bay: Water quality, light regime, and physical-chemical factors. *Estuaries* 27(3):363-377.

King, D.M. and K.J. Adler. 1992. Scientifically defensible compensation ratios for wetland mitigation. Pp. 64-73 in *Effective Mitigation: Mitigation Banks and Joint Projects in the Context of Wetland Management Plans*. Association of State Wetland Managers, Windham, Maine.

Kinlan, B.P. and S.D. Gaines. 2003. Propagule dispersal in marine and terrestrial environments: a community perspective. *Ecology* 84:2007-2020.

Knutson, P.L. and W.W. Woodhouse, Jr. 1983. *Shore stabilization with salt marsh vegetation*. Special Rep. 9. U.S. Army Corps of Engineers, Coastal Engineering Research Center, Fort Belvoir, Virginia.

Koch, E.W. 2001. Beyond light: Physical, geological and geochemical parameters as possible submersed aquatic vegetation habitat requirements. *Estuaries* 24:1-17.

Koch, E.W., J. Ackerman, M. van Keulen and J. Verduin. 2005. Fluid Dynamics in Seagrass Ecology: from Molecules to Ecosystems. Pp. 193-226 in *Biology of Seagrasses*, Larkum, A., R. Orth and C. Duarte (eds.) Springer Verlag, Berlin, Germany.

Komar, P.D. 1998a. *Beach processes and sedimentation*. Prentice Hall, Upper Saddle River, New Jersey.

Komar, P.D. 1998b. *The Pacific Northwest Coast*. Duke University Press, Durham, North Carolina.

Komar, P.D. and D.L. Inman. 1970. Longshore sand transport on beaches. *Journal of Geophysical Research* 75:5914-5927.

Kraus, N.C. and W.C. McDougal. 1996. The effects of seawalls on the beach: Part I, An updated literature review. *Journal of Coastal Research* 12(3):691-701.

Kraus, N.C. and O.H. Pilkey. 1988. The effects of seawalls on the beach. *Journal of Coastal Research*, Special Issue 4:1-28.

Krone, R.B. 1962. *Flume Studies of the Transport of Sediment in Estuarial Processes, Final Report*. University of California, Hydraulic Engineering Laboratory and Sanitary Engineering Research Laboratory, Berkeley, California.

Kunz, K. 2005. *The Regulatory Program*. Presentation to the Committee on Mitigating Shore Erosion along Sheltered Coasts, Seattle, Washington, October 2005. National Research Council, Washington, DC.

Kusler, J.A. and M.E. Kentula. 1989. *Wetland Creation and Restoration: the Status of the Science.* Vol. 1 Regional Review EPA/600/3-89/038. Environmental Protection Agency, Environmental Research Lab, Corvallis, Oregon.

Lafferty, K. 2001. Disturbance to wintering western snowy plovers. *Biological Conservation* 101:315.

Lein, J.K. 2003. *Integrated Environmental Planning.* Blackwell Science, Oxford, United Kingdom.

Levings, C.D., K. Conlin and B. Raymond. 1991. Intertidal habitats used by juvenile chinook salmon (*Oncorhynchus tshawytscha*) rearing in the North Arm of the Fraser River estuary. *Marine Pollution Bulletin* 22:20-26.

Lewis, R.R. III, A.B. Hodgson and G.S. Mauseth. 2005. Project facilitates the natural reseeding of mangrove forests (Florida). *Ecological Restoration* 23:276-277.

Liffmann, M. 1997. Aquatic nuisance species with a focus on zebra mussels. Pp. 91-96 in *Proceedings of the 7th International Zebra Mussel and Aquatic Species Conference*, held on January 1997, in New Orleans, Louisiana. International Conference on Aquatic Invasive Species, Pembroke, Ontario, Canada.

Lincoln Smith, M.P., C.A. Hair and J.D. Bell. 1994. Man-made rock breakwaters as fish habitats: comparisons between breakwaters and natural reefs within an embayment in southeastern Australia. *Bulletin of Marine Science* 55:1344.

Linhart, J., S. Vlockova and V. Uvira. 2002. Bryophytes as special mesohabitat for meiofauna in a rip-rapped channel. *River Research Applications* 18:321-330.

Lubchenco, J., S.R. Palumbi, S.D. Gaines and S. Andelman. 2003. Plugging a hole in the ocean: the emerging science of marine reserves. *Ecological Application* 13:S3-S7.

Maine Department of Environmental Protection. 1998. *Maine Shoreland Zoning: A Handbook For Shoreland Owners.* DEPLW0674-C04. Maine Department of Environmental Protection, Bureau of Land and Water Quality, Augusta. [Online] Available at: http://www.maine.gov/dep/blwq/docstand/szpage.htm [February 14, 2006].

Markley, S.M., G.R. Milano and E. Calas. 1992. Biscayne Bay Restoration and Enhancement Program shoreline and habitat enhancement guide. Pp. 111-120 in *Proceedings of the 19th Annual Conference on Wetlands Restoration and Creation*, Webb, F.J. (ed.). Hillsborough Community College, Plant City, Florida.

Marsh, L.L., D.R. Porter and D.A. Salvesen. 1996. *Mitigation Banking: Theory and Practice.* Island Press, Washington, DC. 225 pp.

Martin, D., F. Bertasi, M.A. Colangelo, M. de Vries, M. Frost, S.J. Hawkins, E. Macpherson, P.S. Moschella, M.P. Satta, R.C. Thompson and V.U. Ceccherelli. 2005. Ecological impact of coastal defense structures on sediment and mobile fauna: evaluating and forecasting consequences of unavoidable modifications of native habitats. *Coastal Engineering* 52:1027-1051.

Maryland Department of Natural Resources. 1992. *Shore Erosion Control Guidelines for Waterfront Property Owners.* Maryland Department of Natural Resources, Water Resources Administration, Annapolis. [Online] Available at: http://www.dnr.state.md.us/grantsandloans/sec_resources.html [November 29, 2005].

Maryland Department of Natural Resources. 2006. *Shore Erosion Control.* Maryland Department of Natural Resources, Annapolis. [Online] Available at: http://www.dnr.state.md.us/grantsandloans/secintro.html [July 20, 2006].

Milano, G.R. 1999. Cape Florida Recreation Area wetlands restoration. Pp. 110-119 in. *Proceedings of the Twenty Fifth Annual Conference on Ecosystems Restoration and Creation*, Cannizzaro, P.J. (ed.). Hillsborough Community College, Tampa, Florida.

Mitsch, W.J. and R.F. Wilson. 1996. Improving the success of wetland creation and restoration with know-how, time, and self-design. *Ecological Applications* 6:77-83.

Mobile Bay National Estuary Program (MBNEP). 2002. *Comprehensive Conservation and Management Plan.* Mobile Bay National Estuary Program, Mobile, Alabama. [Online] Available at: http://www.mobilebaynep.com/Publications.htm [February 14, 2006].

Montgomery, J.A. 2000. The use of natural resource information in wetland ecosystem creation and restoration. *Ecological Applications* 18:45-50.

Morton, R.A., G. Tiling and N.F. Ferina. 2003. Causes of hotspot wetland loss in the Mississippi delta plain. *Environmental Geosciences* 10:71-80.

Moschella, P.S., M. Abbiati, P. Aberg, L. Airoldi, J.M. Anderson, F. Bacchiocchi, F. Bulleri, G.E. Dinesen, M. Frost, E. Gacia, L. Granhag, P.R. Jonson, M.P. Satta, A. Newell and J.A. Ott. 1999. Macrobenthic Communities and Eutrophication. Pp. 265-293 in *Land-Use, Water Quality and Fisheries: A Comparative Ecosystem Analysis of the Northern Adriatic Sea and the Chesapeake Bay*, Malone, T.C., N.A. Smodlaka, A. Malej and L.W. Harding, Jr. (eds). American Geophysical Union, Washington, DC.

Moschella, P.S., M. Abbiati, P. Åberg, L. Airoldi, J.M. Anderson, F. Bacchiocchi, F. Bulleri, G.E. Dinesen, M. Frost, E. Gacia, L. Granhag, P.R. Jonsson, M.P. Satta, A. Sundelöf, R.C. Thompson and S.J. Hawkins. 2005. Low-crested coastal defense structures as artificial habitats for marine life: using ecological criteria in design. *Coastal Engineering* 52:1053-1071.

Myers, D. 2005. *Avoiding the Need for Armoring Marine Shorelines on Sheltered Coasts*. Presentation to the Committee on Mitigating Shore Erosion along Sheltered Coasts, Seattle, Washington, October 2005. National Research Council, Washington, DC.

National Assessment Synthesis Team (NAST). 2000. *Climate Change Impacts on the United States: The Potential Consequences of Climate Variability and Change*. Overview by the National Assessment Synthesis Team for the U.S. Global Change Research Program, Cambridge University Press, Cambridge, United Kingdom.

National Oceanic and Atmospheric Administration (NOAA). 2001. *Tidal Datums and Their Applications*. NOAA Special Publication NOS CO-OPS 1. NOAA, National Ocean Service, Center for Operational Oceanographic Products and Services, Silver Spring, Maryland.

National Oceanic and Atmospheric Administration (NOAA). 2005. *Coastal Zone Management Program*. National Oceanic and Atmospheric Administration, National Ocean Service, Office of Ocean and Coastal Resource Management, Silver Spring, Maryland. [Online] Available at: http://coastalmanagement.noaa.gov/czm/national.html [March 13, 2006].

National Oceanic and Atmospheric Administration (NOAA). 2006. *NOAA Tides and Currents*. National Oceanic and Atmospheric Administration, Center for Operational Oceanographic Products and Services, Silver Spring, Maryland. [Online] Available at: http://tidesandcurrents. noaa.gov/ [April 3, 2006].

National Research Council (NRC). 1990. *Mitigating Coastal Erosion*. National Academy Press, Washington, DC.

National Research Council (NRC). 1992. *Restoration of Aquatic Ecosystems*. National Academy Press, Washington, DC.

National Research Council (NRC). 1995a. *Beach Nourishment and Protection*. National Academy Press, Washington, DC.

National Research Council (NRC). 1995b. *Wetlands: Characteristics and Boundaries*. National Academy Press, Washington, DC.

National Research Council (NRC). 1999. *Sharing the Fish: Toward a National Policy on Individual Fishing Quotas*. National Academy Press, Washington, DC.

National Research Council (NRC). 2001. *Compensating for wetland losses under the clean water act*. National Academy Press, Washington, DC.

National Research Council (NRC). 2003. *Science and the Greater Everglades Ecosystem Restoration: An Assessment of the Critical Ecosystem Studies Initiative*. The National Academies Press, Washington, DC.

National Research Council (NRC). 2004a. *Adaptive Management for Water Resources Project Planning*. The National Academies Press, Washington, DC.

National Research Council (NRC). 2004b. *River Basins and Coastal Systems Planning Within the U.S. Army Corps of Engineers*. The National Academies Press, Washington, DC.

National Research Council (NRC). 2005. *Valuing Ecosystem Services.* The National Academies Press, Washington, DC.

National Research Council (NRC). 2006. *Drawing Louisiana's New Map: Addressing Land Loss in Coastal Louisiana.* The National Academies Press, Washington, DC.

National Research Council, Board on Natural Disasters. 1999. Mitigation emerges as major strategy for reducing losses caused by natural disasters. *Science* 284:1943-1947.

Nature. 1998. Audacious bid to value the planet whips up a storm. *Nature* 395:430. [Online] Available at: http://fire.biol.wwu.edu/hooper/costanza97naturecomment.pdf [December 12, 2005].

New York Sea Grant. 1984. *Analysis, Design and Construction of Coastal Structures.* Geotechnical Engineering Group, Cornell University, for New York Sea Grant Institute. New York Sea Grant, Stony Brook, New York.

Newell, R.I.E. and E.W. Koch. 2004. Modeling seagrass density and distribution in response to changes in turbidity stemming from bivalve filtration and seagrass sediment stabilization. *Estuaries* 27:793-806.

Newell, R.I.E. and J. Ott. 1999. Macrobenthic Communities and Eutrophication. Pp. 265-293 in *Ecosystems at the Land-Sea Margin: Drainage Basin to Coastal Sea. Coastal and Estuarine Studies*, Malone, T.C., A. Malej, L.W. Harding, Jr., N. Smodlaka and R.E. Turner (eds). American Geophysical Union, Washington, DC.

Nielsen, P. 1992. *Coastal Bottom Boundary Layers and Sediment Transport.* World Scientific Publishing Company, River Edge, New Jersey.

Nordstrom, K.F. 1992. *Estuarine Beaches; An Introduction to the Physical and Human Factors Affecting Use and Management of Beaches in Estuaries, Lagoons, Bays and Fjords*, Elsevier Applied Science, London, United Kingdom.

Nordstrom, K. 2005. Beach nourishment and coastal habitats: Research deeds to improve compatibility. *Restoration Ecology* 13(1):215-222.

Northwest Regional Planning Commission. *The Shoreline Stabilization Handbook for Lake Champlain and Other Inland Lakes.* Northwest Regional Planning Commission, St. Albans, Vermont. [Online] Available at: http://nsgd.gso.uri.edu/lcsg/lcsgh04001.pdf [February 28, 2006].

Ogden, J. and P. Lobel. 1978 The role of herbivorous fishes and urchins in coral reef communities. *Environmental Biology of Fishes* 3:49-63.

Orth, R.J., M. Luckenbach and K.A. Moore. 1994. Seed dispersal in a marine macrophyte: Implications for colonization and restoration. *Ecology* 75:1927-1939.

Orth, R.J., M.C. Harwell, E.M. Bailey, A. Bartholomew, J.T. Jawad, A.V. Lombana, K.A. Moore, J.M. Rhode and H.E. Woods. 2000. A review of issues in seagrass seed dormancy and germination: implications for conservation and restoration. *Marine Ecology Progress Series* 200:277-288.

Packham, J.R. and A.J. Willis. 1997. *Ecology of Dunes, Saltmarsh and Shingle.* Chapman & Hall, London.

Peterson, C.H. and M.J. Bishop. 2005. Assessing the environmental impacts of beach renourishment. *BioScience* 55:887-896.

Pethick, J.S. 1996. The geomorphology of a mudflat. Pp. 41-62 in *Estuarine Shores: Evolution, Environment and Human Health*, Nordstrom, K.F. and C.T. Roman (eds). Cambridge University Press, Cambridge, United Kingdom.

Pettijohn, F.J., P.E. Potter and R. Siever. 1973. *Sand and Sandstone.* Springer-Verlag, New York, New York.

Pile Buck. 1990. *Coastal Construction.* Pile Buck, Incorporated, Jupiter, Florida.

Pilkey, O.H. and K.L. Dixon. 1996. *The Corps and the Shore.* Island Press, Washington, DC.

Pilkey, O. and H. Wright. 1988. Seawalls vs. beaches. *Journal of Coastal Research* SI4:41-64.

Quigley, J.T. and D.J. Harper (eds.). 2004. *Stream bank protection with rip-rap: An evaluation of the effects on fish habitat.* Canadian Manuscript Report of Fisheries and Aquatic Sciences Report no. 2701. Canada Department of Fisheries and Oceans, Ottawa, Ontario, Canada. 76 pp.

Radomski, P. and T.J. Goeman. 2001. Consequences of Human Lakeshore Development on Emergent and Floating-Leaf Vegetation Abundance. *North American Journal of Fisheries Management* 21:46-61.

Rasheed, M., M.I. Badran and M. Huettel. 2003. Particulate matter filtration and seasonal nutrient dynamics in permeable carbonate and silicate sands of the Gulf of Aqaba, Red Sea. *Coral Reefs* 22:167-177.

Rennie, T.H. 1990. Using new work and maintenance material for marsh creation in the Galveston District. Beneficial Uses of Dredged Material. Pp. 184-187 in *Proceedings of the Gulf Coast Regional Workshop*, held on April 26-28, 1988 in Galveston, Texas. Technical Report D-90-3. U.S. Army Corps of Engineers, Waterways Experiment Station, Vicksburg, Mississippi.

Rice, D.W., T.A. Dean, F.R. Jacobsen and A.M. Barnett. 1989. Transplanting giant kelp *Macrocystis pyrifera* in Los Angeles Harbor: productivity of the kelp population. *Bulletin of Marine Science* 44:1070

Riggs, S.R. 2001. *The Sound Front Series: Shoreline Erosion in North Carolina Estuaries*. North Carolina Sea Grant, Raleigh, North Carolina.

Rogers, S. 2005. *Complexities in Evaluating the Impact of Estuarine Erosion Management Alternatives*. Presentation to NRC Committee on Mitigating Shore Erosion along Sheltered Coasts, Washington, DC, June 2005. National Research Council, Washington, DC.

Rogers, S. and T.E. Skrabal. 2001. *Managing Erosion on Estuarine Shorelines*. North Carolina Sea Grant, Raleigh.

Ruddy, G., C.M. Turley and T.E.R. Jones. 1998. Ecological interaction and sediment transport on an intertidal mudflat. I. Evidence for a biologically mediated sediment-water interface. Pp. 135-148 in *Sedimentary processes in the intertidal zone*, Black K.S., D.M. Paterson, A. Cramp (eds). Geological Society of London, United Kingdom.

Ryer, C.H., J. van Montfrans and R.J. Orth. 1990. Utilization of a seagrass meadow and tidal marsh creek by blue crabs *Callinectes sapidus*: Spatial and temporal patterns of molting. *Bulletin of Marine Science* 46:95-104.

Schenk, E.R. and N.B. Rybicki. 2006. Exploring causes of a seagrass transplant failure in the Potomac River (Virginia). *Ecological Restoration* 24(2):116-118.

Sebens, K.P. 1996. Biodiversity of coral reefs: what are we losing and why? *Biological Conservation* 76(2):210.

Seitz, A.C., B.L. Norcross, D. Wilson and J.L. Nielsen. 2005. Identifying spawning behavior in Pacific halibut, *Hippoglossus stenolepis*, using electronic tags. *Environmental Biology of Fishes* 73:445-451.

Seitz, R.D., R.N. Lipcius, N.H. Olmstead, M.S. Seebo and D.M Lambert. 2005. Influence of shallow-water habitats and shoreline development upon abundance, biomass, and diversity of benthic prey and predators in Chesapeake Bay. *Marine Ecology Progress Series* 326:11-27.

Shields, A. 1936. Anwendung der Aehnlichkeitsmechanik und turbulenzforchung auf die geschiebebewegung [German.] Ph.D. Thesis, Mitt. Preuss Ver.-Anst., Berlin, Germany 26 pp. [Online] Available at: http://www.waterbouw.tudelft.nl/public/verhagen/ [March 2, 2006].

Shipman, H. and J. Parsons. 2005. *Shoreline erosion on Puget Sound: Field trip to West Seattle and Southwest King County, including Lincoln and Seahurst Parks*. Presentation to the Committee on Mitigating Shore Erosion along Sheltered Coasts, Seattle Washington, October 2005. National Research Council, Washington, DC.

Simenstad, C.A. and R.M. Thom. 1996. Functional equivalency trajectories of the restored Gog-Le-Hi-Te estuarine wetland. *Ecological Applications* 6:38-56.

Snedaker, S.C. and P.D. Biber. 1997. Restauración de manglares en los Estados Unidos de América: Estudio de Caso de la Florida. Pp. 187-208 in *La Restauración de Ecosistemas de Manglares*, Field. C. (ed.). [Spanish]. Editora de Arte, Managua, Nicaragua.

Spalding, V.L. and N.L. Jackson. 2001. Field investigation of the influence of bulkheads on meiofaunal abundance in the foreshore of an estuarine sand beach. *Journal of Coastal Research* 17:363-370.

Spalding, M., M. Taylor, C. Ravilious, F. Short and E. Green. 2003. The distribution and status of seagrasses. Pp. 5-26 in *World Atlas of Seagrasses*, Green, E.P. and F.T. Short (eds). University of California Press, Berkeley.

Spaling, H. and B. Smit. 1993. Cumulative environmental change: Conceptual frameworks, evaluation approaches, and institutional perspectives. *Environmental Management* 17:587-600.

State of Maryland Shore Task Force. 2000. *Final Report*. Maryland Department of Natural Resources, Annapolis.

Stein, E.D. 1999. Mitigation banking: Challenges and lessons learned. *Society of Wetland Scientists Bulletin* 16(3):18-21.

Stephens, J. and D. Pondella. 2002. Larval productivity of a mature artificial reef: the ichthyoplankton of King Harbor, California, 1974-1997. *ICES Journal of Marine Science* 59:S51-S58.

Stephens, J., P.A. Morris, D. Pondella, T.A. Koonce and G.A. Jordan. 1994. Overview of the dynamics of an urban artificial reef fish assemblage in King Harbor, California, USA, 1974-1991: a recruitment driven system. *Bulletin of Marine Science* 55:1224-1239.

Stoddart, D.R. 1974. Post-hurricane changes on the British Honduras reefs: re-survey of 1972. Pp. 473-484 in *Proceedings of the Second International Symposium on Coral Reefs*, Vol. 2. The Great Barrier Reef Committee, Brisbane, Australia.

Taggert, B.E. and M.L. Schwartz. 1988. Net shore-drift Direction Determination: A Systematic Approach. *Journal of Shoreline Management* 3:285-309.

Tait, J.F. and G.B. Griggs. 1990. Beach response to the presence of a seawall: a comparison of field observations. *Shore and Beach* 58(2):11-28.

Tanski, J. 2005. *Shore Hardening on Sheltered Coasts: Impacts and Issues*. Presentation to the Committee on Mitigating Shore Erosion along Sheltered Coasts, meeting held on October 4-6, 2005 in Seattle, Washington. National Research Council, Washington, DC.

Teal, J. and L. Weishar. 2005. Ecological engineering, adaptive management and restoration management in Delaware Bay salt marsh restoration. *Ecological Engineering* 25:304-314.

Thom, R.M., D.K. Shreffler and K. Macdonald. 1994. *Shoreline Armoring Effects on Coastal Ecology and Biological Resources in Puget Sound, Washington. Coastal Erosion Management Studies Volume 7*. Shorelands and Coastal Zone Management Program, Washington Department of Ecology, Olympia.

Titus, J.G. and V. Narayanan. 1995. *The Probability of sea level rise*. U.S. Environmental Protection Agency, Washington, DC. 186 pp.

Toft, J., C. Simenstad, J. Cordell and L. Stamatiou. 2004. *Fish distribution, abundance and behavior in nearshore habitats along city of Seattle marine shorelines, with emphasis on juvenile Salmon*. Technical Report no. 0401. Washington University, Fishery Research Institute, School of Aquatic Fishery Science, Seattle.

Trono, K.L. 2003. An Analysis of the Current Shoreline Management Framework in Virginia: Focus on the Need for Improved Agency Coordination. Internship Report. University of Miami, Rosenstiel School of Marine and Atmospheric Science, Miami, Florida.

Turner, I.L. and G. Masselink. 1998. Swash infiltration-exfiltration and sediment transport. *Journal of Geophysical Research* 103:30813-30824.

U.S. Army Corps of Engineers (USACE). 1971. *National Shoreline Study—California Regional Inventory: San Francisco, California*. U.S. Army Corps of Engineers, San Francisco District, California. 105 pp.

U.S. Army Corps of Engineers (USACE). 1981. *Low Cost Shore Protection: A Guide for Engineers and Contractors*. GAI Consultants, Monroeville, Pennsylvania.

U.S. Army Corps of Engineers (USACE). 1984. *Shore Protection Manual*. (4th ed.)(2). U.S. Army Corps of Engineers, Washington, DC.

U.S. Army Corps of Engineers (USACE). 1995. *Shore Protection and Beach Erosion Control Study: Economic Effects of Induced Development in Corps-Protected Beachfront Communities*. IWR Report 95-PS-1. U.S. Army Corps of Engineering, Water Resources Support Center, Alexandria, Virginia.

U.S. Army Corps of Engineers (USACE). 1996. *Shoreline Protection and Beach Erosion Control Study: Final Report: An Analysis of the U.S. Army Corps of Engineers Shore Protection Program.* (IWR Report 96-PS-1). U.S. Army Corps of Engineering, Water Resources Support Center, Alexandria, Virginia.

U.S. Army Corps of Engineers (USACE). 2000. *Coastal Engineering Manual.* U.S. Army Corps of Engineers, Coastal and Hydraulics Laboratory, Vicksburg, Mississippi. [Online] Available at: http://chl.erdc.usace.army.mil/CHL.aspx?p=s&a=Publications;8 [March 3, 2006].

U.S. Army Corps of Engineers (USACE). 2001. *Permitting and Appeals.* U.S. Army Corps of Engineers, Washington, DC [Online] Available at: http://www.saj.usace.army.mil/permit/permitting/permitting.htm [December 1, 2005].

U.S. Army Corps of Engineers (USACE). 2002a. *Mitigation Plan Development.* U.S. Army Corps of Engineers, Washington, DC [Online] Available at: http://www.saw.usace.army.mil/wetlands/Mitigation/mitplan.html [March 28, 2006].

U.S. Army Corps of Engineers (USACE). 2002b. Coastal Engineering Manual. EM 1110-2-1100. U.S. Army Corps of Engineers, Washington, DC [Online] Available at: http://chl.erdc.usace.army.mil/CHL.aspx?p=s&a=Publications;8 [October 4, 2006].

U.S. Army Corps of Engineers (USACE). 2004. *Mitigation and Monitoring Proposal Guidelines: San Francisco and Sacramento Districts.* U.S. Army Corps of Engineers, San Francisco, California. [Online] Available at: http://www.spn.usace.army.mil/regulatory/ [March 3, 2006].

U.S. Army Corps of Engineers (USACE). 2005. *Regional Sediment Management.* U.S. Army Corps of Engineers, Washington, DC. [Online] Available at: http://www.wes.army.mil/rsm/ [March 13, 2006].

U.S. Army Corps of Engineers (USACE). 2006a. *Coastal and Hydraulics Laboratory Projects.* U.S. Army Corps of Engineers, Coastal and Hydraulics Laboratory, Vicksburg, Mississippi. [Online] Available at: http://chl.erdc.usace.army.mil/chl.aspx?p=i&a=Projects!0 [April 3, 2006].

U.S. Army Corps of Engineers (USACE). 2006b. *Compensatory Mitigation Practices in the U.S. Army Corps of Engineers.* USACE Institute for Water Resources Working Paper prepared by Martin, Brumbaugh, Scodari and Olson. U.S. Army Corps of Engineers, Institute for Water Resources, Alexandria, Virginia.

U.S. Army Corps of Engineers (USACE), Environmental Protection Agency, Fish and Wildlife Service, Natural Resources Conservation Service, National Marine Fisheries Service, Massachusetts Executive Office of Environmental Affairs, Massachusetts Department of Environmental Protection and Massachusetts Coastal Zone Management. 1998. *Massachusetts Wetlands Restoration and Banking Program. Pilot Wetlands Banking Project: Memorandum of Agreement.* U.S. Army Corps of Engineers, North Atlantic District, Brooklyn, New York. [Online] Available at: http://www2.eli.org/wmb/umbrelladetail.cfm?AgreementID=21 [March 3, 2006].

U.S. Army Corps of Engineers (USACE), State of Maryland and Commonwealth of Virginia. 1990. *Chesapeake Bay Shoreline Erosion Study.* U.S. Army Corps of Engineers, Baltimore District, Maryland.

U.S. Commission on Ocean Policy. 2004. *An Ocean Blueprint for the 21st Century.* Final Report. U.S. Commission on Ocean Policy, Washington, DC.

U.S. Department of Housing and Urban Development (USDHUD). 1992. *Flood Insurance Study, Town of Cape Charles.* Report 510106. U.S. Department of Housing and Urban Development, Federal Insurance Administration, Flood Insurance Study, Cape Charles, Northampton County, Virginia.

U.S Environmental Protection Agency. 2000. *Sea Level.* U.S Environmental Protection Agency, Washington, DC. [Online] Available at: http://yosemite.epa.gov/oar/globalwarming.nsf/content/ClimateTrendsSeaLevel.html [September 18, 2006].

U.S. Geological Survey (USGS). 1999. *U.S. Geological Survey's Priority Ecosystems Science Initiative: Priorities Program.* U.S. Geological Survey, Reston, Virginia. [Online] Available at: http://access.usgs.gov/prog_priorities.html#top [March 28, 2006].

U.S. Geological Survey (USGS). 2002. *Ecological Research on Wetlands and Submersed Aquatic Vegetation*. U.S. Geological Survey, Water Resources Discipline, Reston, Virginia. [Online] Available at: http://water.usgs.gov/nrp/proj.bib/sav/wethome.htm [September 7, 2006].

Van Aarde, R.J., S.M. Ferreira, J.J. Kritzinger, P.J. Van Dyk, M. Vogt and T.D. Wassenaar. 1996. An evaluation of habitat rehabilitation on coastal dune forests in northern KwaZulu/Natal, South Africa. *Restoration Ecology* 4(4):334-345.

van Koningsveld, M. and J.P.M. Mulderc. 2004. Sustainable Coastal Policy Developments in The Netherlands: A Systematic Approach Revealed. *Journal of Coastal Research* 20(2):375-385.

Virginia Institute of Marine Science. 2005. *Interagency Shoreline Management Consensus Document*. Submitted by the Center for Coastal Resources Management, Virginia Institute of Marine Science and submitted to the Virginia Coastal Program, Department of Environmental Quality. Virginia Institute of Marine Science, Gloucester Point, Virginia.

Virginia Institute of Marine Science. 2006. *Chesapeake Bay Breakwater Database*. Virginia Institute of Marine Science, Gloucester Point. [Online] Available at: http://www2.vims.edu/breakwater/ [February 8, 2006].

Virginia Marine Resources Commission. 1989. *Shoreline Development BMP's*. Virginia Marine Resources Commission, Newport News.

Walters, C.J. 1986. *Adaptive Management of Renewable Resources*. John Wiley and Sons, Chichester, United Kingdom.

Ward, L.G., P.S. Rosen, W.J. Neal, O.H. Pilkey, Jr., O.H. Pilkey, Sr., G.L. Anderson and S.J. Howie. 1989. *Living with Chesapeake Bay and Virginia's Ocean Shores*. Duke University Press, Durham, North Carolina.

Washington State Department of Ecology. 2006. *Coastal Zone Management*. Washington State Department of Ecology, Seattle. [Online] Available at: http://www.ecy.wa.gov/programs/sea/czm/index.html [March 13, 2006].

Weinstein, M.P., J.H. Balletto, J.M. Teal and D.F. Ludwig. 1997. Success criteria and adaptive management for a large-scale wetland restoration project. *Wetlands Ecology and Management* 4:111-127.

Weisberg, S.B. and V.A. Lotrich. 1982. The importance of an infrequently flooded interitidal marsh surface as an energy source for the mummichog, *Fundulus heteroclitus*: An experimental approach. *Marine Biology* 66:307-310.

Weishar, L., J. Teal and J. Balletto. 1998. The role of adaptive management in the restoration of degraded diked salt hay farm wetlands. Pp. 356-361 in *Proceedings of the American Society of Civil Engineering Wetlands Engineering and River Restoration Conference*, Denver, Colorado, March 22-27, 1998. American Society of Civil Engineering, Reston, Virginia.

Wicks, E.C. 2005. The effect of sea level rise on seagrasses: is sediment adjacent to retreating marshes suitable for seagrass growth? MS Thesis, University of Maryland, College Park. 140 pp.

Woodford, J.E. and M.W. Meyer. 2003. Impact of lakeshore development on green frog abundance. *Biological Conservation* 110:277-284.

Wright, L.D., J. Chappell, B.G. Thom, M.P. Bradshaw and P. Cowell. 1979. Morphodynamics of reflective and dissipative beach and inshore systems: southeastern Australia. *Marine Geology* 32:105-140.

Wright, L.D., A.D. Short, J.D. Boon, B. Hayden, S. Skimball and J.H. List. 1987. The morphodynamic effects of incident wave groupiness and tide range on an energetic beach. *Marine Geology* 74:1-20.

Zakaia, Z. and N.E. Chadwick-Furman. 2002. Impacts of intensive recreational diving on reef corals at Eilat, northern Red Sea. *Biological Conservation* 105:179-187.

Zanuttigh, B., L. Martinelli, A. Lamberti, P. Moschella, S. Hawkins, S. Marzetti and V. Ugo Ceccherelli. 2005. Environmental design of coastal defense in Lido di Dante, Italy. *Coastal Engineering* 52:1089-1125.

Zedler, J.B. 1996. Ecological issues in wetland mitigation: an introduction to the forum. *Ecological Applications* 6:33-37.

Zedler, J.B. 1998. Replacing endangered species habitat: the acid test of wetland ecology. Pp. 364-379 in *Conservation Biology for the Coming Age*, Fiedler, P.L. and P.M. Kareiva (eds.) Chapman and Hall, New York, New York.

Zhao, B., U. Kreuter, B. Li, Z. Ma, J. Chen and N. Nakagoshi. 2004. An ecosystem service value assessment of land use change on Chongming Island, China. *Land Use Policy* 21:139-148.

Appendixes

A

Committee and Staff Biographies

COMMITTEE

Jeff Benoit (*Chair*) earned his M.S. in coastal geology in 1978 from Georgia Institute of Technology/Skidaway Institute of Oceanography. He is the Director of Coastal and Marine Programs for SRA International. Prior to joining SRA in 2005 he was the Proprietor of J.R. Benoit Consulting. Mr. Benoit's professional interests focus on coastal management, marine conservation, policy analysis, program assessment, and coastal hazard mitigation planning. He previously served as Director of the National Oceanic and Atmospheric Administration's (NOAA) Office of Ocean and Coastal Resource Management (OCRM) from 1993 to 2001. He was chair of NOAA's Sustain Healthy Coasts strategic planning team from 1994 to 1997. Mr. Benoit also chaired one of five Working Groups established by the U.S. Coral Reef Task Force and coordinated drafting the National Action Plan to Conserve Coral Reefs. In his efforts to further global marine issues, Mr. Beniot has been involved with Integrated Coastal Management activities in China, Ireland, Jordan, Indonesia, and many Pacific and Caribbean Island communities. Prior to accepting his position at OCRM, he served as Coastal Geologist and later Director of the Massachusetts Coastal Zone Management Program. Mr. Benoit has served on the National Research Council's committee on Restoration and Protection of Coastal Louisiana.

C. Scott Hardaway received his M.S. in 1980 from the Department of Geology, East Carolina University. He is a Marine Scientist Supervisor for the Physical Sciences Department of the Virginia Institute of Marine Science at the College of William and Mary. Mr. Hardaway is primarily involved in process and response of shoreline systems, which includes geomorphology of coastal change, wave mechanics and sediment transport, numerical modeling of hydrodynamic processes, shoreline change and shore protection, particularly breakwater and beach fill performance in Chesapeake Bay. He has worked on restoring and protecting wetlands from damage due to erosion and sea level rise in the Chesapeake Bay region. Mr. Hardaway has also contributed to the development of the U.S. Army Corps of Engineers' (Norfolk District) ecosystem restoration projects to protect and restore wetlands. These projects include areas such as Tangier Island and Saxis Island, Virginia. His professional memberships include the Virginia Board of Geology, North Carolina Board for Licensing of Geologists, and the American Society of Civil Engineers.

Debra Hernandez received her M.S. in civil engineering from Clemson University in 1987. She is a professional engineer and her background is in coastal management and engineering. Ms. Hernandez is currently the President of Hernandez and Company, and worked previously as Director of Program and Policy Development for the South Carolina Department of Health and Environmental Control. She is an acknowledged leader in coastal management with 18 years of experience and extensive policy, legislative, and technical expertise. Ms. Hernandez's expertise lies in federal and state coastal and environmental management laws, regulations and policies. She currently serves on the National Academies' Ocean Studies Board and the Ocean Research and Resources Advisory Panel. Ms. Hernandez is also a founding board member and current vice-chair for the Coastal States Stewardship Foundation, whose purpose is to support healthy coasts and vibrant coastal communities. Additionally, she chaired the Coastal States Organization (CSO) from 2002 to 2004. CSO represents the interests of 35 governors from coastal states on federal activities relating to coastal management.

Robert Holman earned his Ph.D. from Dalhousie University (Physical Oceanography) in 1979 and his B.S. in honors mathematics and physics from the Royal Military College at Kingston Canada in 1972. He is Professor at the College of Oceanic Atmospheric Sciences at Oregon State University. Dr. Holman's research interests include beach processes; measurements of nearshore waves and currents; models of sandbar generation and morphology; application of remote sensing to nearshore processes, large-scale coastal behavior. He is currently performing research projects on the interactions of waves with nearshore morphology; measurement and modeling of sandbar morphology climatology; remote-sensing signatures in the nearshore. These are being completed for the Coastal Imaging Laboratory at Oregon State University, a laboratory that Dr. Holman has been developing since he joined Oregon State University in 1979. Dr. Holman has published more than 50 refereed papers encompassing theoretical, observation, and technical areas including recent papers on pattern formation in the nearshore; the statistics and kinematics of transverse sandbars on an open coast; and the role of morphological feedback in surf zone sand bar response. He was recently appointed to a four-year term as the FY 2004 Secretary of the Navy/Chief of Naval Research Chair in Oceanographic Science. Dr. Holman has also served as a consultant for Navy Special Projects at Mitre Corporation; the National Science Foundation ad hoc committee for establishment of a coastal engineering program; the NSF Coastal Geology task force; several National Research Council committees; and many other professional and technical organizations.

Evamaria Koch earned her Ph.D. in Marine Science from the University of South Florida in 1993. She is an Associate Professor at Horn Point Laboratory at the University of Maryland Center for Environmental Science. Dr. Koch's research and areas of professional expertise include seagrass ecology, hydrodynamically-mediated processes in meadows of aquatic plants, seagrass habitat engineering, global changes in coastal plant systems, culture and micropropagation of marine macrophytes, and ecophysiology of marine plants. She has recently published on fluid dynamics in seagrass ecology, modeling seagrass density and distribution in response to changes in turbidity stemming from bivalve filtration and seagrass sediment stabilization, and habitat requirements for submerged aquatic vegetation in Chesapeake Bay. Dr. Koch's research currently focuses on designing seagrass friendly structures to mitigate shoreline erosion and on the effect of global changes on seagrass communities.

Neil McLellan holds a B.S. in civil engineering from the University of Texas and an M.S. in ocean engineering from Texas A&M University. He is a Senior Project Manager with Shiner Moseley and Associates. Mr. McLellan has been working in the Coastal Engineering field since his graduation from the University of Texas in 1981. His background includes research, detail design, project management, environmental enhancements, and public outreach for projects involving coastal engineering, shoreline protection, habitat creation, navigation, dredging, and dredged material placement. He has worked in the public and private sectors and has published over 40 technical papers on dredging, dredged material placement, wetlands creation, and coastal processes. Mr. McLellan has completed projects on shoreline protection along the Gulf of Mexico in Cameron Parish Louisiana; evaluated options, developed implementation, monitoring and design for innovative shoreline protection along the eastern Gulf Coast of Texas; served as the design engineer for Confined Disposal Facility to receive contaminated materials from a section of the Houston Ship Channel; was the Project Manager for determining feasibility of moving existing bayou from within operating chemical plant and locating outside the plant facility; and provided project management for development of detailed environmental assessment for innovative shoreline protection in Jefferson County, Texas. He is a registered professional engineer and currently serves as a member of the American Society of Civil Engineers, the Texas Shore and Beach Association, and the Western Dredging Association.

Susan Peterson earned her Ph.D. in anthropology from the University of Hawaii. She currently works as a partner in the family consulting business Teal Partners, which she operates with her husband, John Teal. Dr. Peterson served at Woods Hole Oceanographic Institution for over 10 years, specializing in ocean policy, and several years at Boston University in the same role. Most of her work was quantifying and qualifying commercial and recreational fishing effort so that the cultural, social, and economic motivations for fishing could be incorporated into fishery management decisions. In the mid-1980s, she left academia for industry and started Ecological Engineering Associates (EEA) that uses natural systems for wastewater treatment for private, industrial, and municipal customers. She retired from EEA in 1998 to partner her consulting firm. Dr. Peterson's areas of expertise include wastewater treatment technology and policy; municipal and private finance for infrastructure, fisheries management (international and domestic); fisheries and aquaculture development and marketing; coastal zone management (international and domestic); ocean pollution effects on human health, coastal communities and industry; fishing communities, industry structure and infrastructure, fish pricing and markets, decision-making. She serves on the Board of Directors of the Wildlands Trust of Southeastern MA and is vice chair of the regional planning agency for southeastern Massachusetts. She also serves on the Greater New Bedford Workforce Investment Board. Dr. Peterson is also involved with and serves on committees of numerous regional and national conservation organizations.

Denise Reed holds a Ph.D. in geography from the University of Cambridge. She is currently a Professor at the University of New Orleans. Dr. Reed's research focuses on various aspects of sediment dynamics in coastal wetlands, with emphasis on sediment mobilization and marsh hydrology, both natural and altered, as factors controlling sediment deposition. She has participated in numerous research projects concerning marsh and estuarine sediment dynamics on the Gulf and Pacific coasts of the United States as well as in Europe and South America. Dr. Reed has also worked closely with the development of restoration plans in for coastal Louisiana for the last 15 years, being involved in incorporating science into efforts under the Coastal Wetlands Planning, Protection and Restoration Act (CWPPRA) and more recently the Louisiana Coastal Area study. Dr. Reed has recently completed publications on altered hydrology effects on Louisiana salt marsh function and restoration of tidal wetlands in the Sacramento-San Joaquin delta. She currently serves on scientific advisory boards for ecosystem restoration in San Francisco Bay and Jamaica Bay, NY, as well as the CALFED program and the U.S. Army Corps of Engineers.

Daniel Suman holds a J.D. and Certificate in Environmental Law from the University of California at Berkeley and a Certificate of Latin American Studies from Columbia University. He received his Ph.D. in Chemical Oceanography from Scripps Institution of Oceanography, University of California at San Diego. Dr. Suman is currently Professor of Marine Affairs and Policy at the Rosenstiel School of Marine and Atmospheric Science (RSMAS), University of Miami, as well as an Adjunct Professor in the School of Law. He teaches courses in Coastal Law, Environmental Law, Environmental Planning, and Coastal Management. Dr. Suman conducts research on the adaptability of the fishing sectors in Chile, Peru, and Ecuador to ENSO ("El Niño") climate variability. This research involves analyses of changing positions of industrial fishing companies and labor unions, artisanal fishing unions, and government regulators in light of environmental uncertainty. He also has long-standing research interests in mangrove management in Latin American and Caribbean countries. Dr. Suman has organized international training workshops for mangrove managers and analyzes mangrove-related legislation in the Americas. His practical experience in Integrated Coastal Management extends from Panama to Brazil, Ururguay, Ecuador, Mexico, and Italy. Dr. Suman has served as the Articles Editor of the Ecology Law Quarterly and has been a fellow at the Smithsonian Tropical Research Institute. He is currently a member or the IUCN Commission on Environmental Law, the Marine Protected Areas Federal Advisory Committee, as well as the National Research Council's Ocean Studies Board.

STAFF

Susan Roberts (*Study Director*) became the Director of the Ocean Studies Board in April 2004. Dr. Roberts received her Ph.D. in Marine Biology from the Scripps Institution of Oceanography. She worked as a research scientist at the University of California, Berkeley and as a senior staff fellow at the National Institutes of Health. Dr. Roberts' past research experience has included deep sea biology, developmental cell biology, and cancer cytogenetics. She has directed a number of studies for the Ocean Studies Board including *Nonnative Oysters in the Chesapeake Bay* (2004); *Decline of the Steller Sea Lion in Alaskan Waters: Untangling Food Webs and Fishing Nets* (2003); *Effects of Trawling & Dredging on Seafloor Habitat* (2002); *Marine Protected Areas: Tools for Sustaining Ocean Ecosystems* (2001); *Under the Weather: Climate, Ecosystems, and Infectious Disease* (2001); *Bridging Boundaries Through Regional Marine Research* (2000); and *From Monsoons to Microbes: Understanding the Ocean's Role in Human Health* (1999). Dr. Roberts specializes in the science and management of living marine resources.

Sarah Capote earned her B.A. in history from the University of Wisconsin-Madison in the winter of 2001. She is a senior program assistant with the Ocean Studies Board. During her tenure with the Board, Ms. Capote worked on the following reports: *Exploration of the Seas: Voyage into the Unknown* (2003), *Nonnative Oysters in the Chesapeake Bay* (2004), *Future Needs in Deep Sub-*

mergence Science: Occupied and Unoccupied Vehicles in Basic Ocean Research (2004), the interim report for *Elements of a Science Plan for the North Pacific Research Board* (2004), *A Vision for the International Polar Year 2007-2008* (2004), *Marine Mammal Populations and Ocean Noise: Determining When Noise Causes Biologically Significant Effects* (2005), *Final Comments on the Science Plan for the North Pacific Research Board* (2005), *Oil Spill Dispersants: Efficacy and Effects* (2005), *Managing Coal Combustion Residues in Mines* (2006), and *A Review of the Draft Ocean Research Priorities Plan: Charting the Course for Ocean Science in the United States* (2006).

B

Acronyms

CAMA	Coastal Area Management Act
CEM	USACE Coastal Engineering Manual
CICEET	Cooperative Institute for Coastal and Estuarine Environmental Technology
CWA	Clean Water Act
CZMA	Coastal Zone Management Act
EFH	Essential Fish Habitat
EPA	Environmental Protection Agency
ESA	Engendered Species Act
FDEP	Florida Department of Environmental Protection
FEMA	Federal Emergency Management Agency
FIRM	flood insurance rate map
FR	Federal Register
FWCA	Fish and Wildlife Coordination Act
FWPCA	Federal Water Pollution Control Act
FWS	Fish and Wildlife Service
IPCC	Intergovernmental Panel on Climate Change
LCS	low-crested structures
MCL	momentary coastline

MHW   mean high water
MLLW   Mean Lower Low Water
MLWL   mean low-tide water line

NAP   Normaal Amsterdams Peil [Dutch reference for sea level]
NCDENR   North Carolina Department of Environmental and Natural
          Resources
NEPA   National Environmental Policy Act
NHPA   National Historic Preservation Act
NOAA   National Oceanic and Atmospheric Administration
NRC   National Research Council
NRCS   Natural Resource and Conservation Service
NWP   Nationwide Permit

PTD   Public Trust Doctrine

RSM   Regional Sediment Management

SAV   submerged aquatic vegetation
SCDHEC   South Carolina Department of Health and Environmental Control
SPGP   State Programmatic General Permit

USACE   U.S. Army Corps of Engineers
USDHUD   U.S. Department of Housing and Urban Development
USGS   U.S. Geological Survey

VIMS   Virginia Institute of Marine Science

C

Workshop Agenda and Participants List

Committee on Mitigating Shore Erosion along Sheltered Coasts
Workshop
Talaris Conference Center
4000 NE 41st Street
Seattle, WA 98105
October 4-6, 2005
AGENDA

TUESDAY, OCTOBER 4

Open Session

8:00 a.m.   BREAKFAST

PLENARY SESSION #1—Maple Room

8:30 a.m.   Welcome, Introduction, and Purpose of Workshop—
                 Jeff Benoit, *Chair,* **and Sue Roberts**, *Study Director*

8:45 a.m.   How bad is the problem: Case study of Mobile Bay—
                 Scott Douglas, *University of South Alabama*

9:10 a.m.   Permitting process, federal requirements—
                 Kathleen Kunz, *U.S. Army Corps of Engineers*

9:35 a.m.   DISCUSSION

9:50 a.m.   Instructions for the breakout sessions—
               Jeff Benoit, *Chair*

10:00 a.m.   BREAK

CONCURRENT SESSIONS—GEOMORPHIC SETTINGS:
A. Beaches
B. Mudflats and Marshes
C. Bluffs

10:20 a.m.   Introductions and Description of the Breakout Process

10:45 a.m.   Overview presentation by invited speaker
               Beaches Overview:
               Neville Reynolds, *Vanasse Hangen Brustlin, Inc.*
               Mudflats and Marshes Overview:
               John Teal, *Teal Partners*
               Bluffs Overview:
               Hugh Shipman, *Washington Department of Ecology*

11:15 a.m.   Mini-presentations by willing participants

               Discussion of setting-specific geomorphic processes

Noon   LUNCH

1:00 p.m.   Facilitated discussion of erosion:
               • causes,
               • how it works,
               • which kinds of substrates are sensitive,
               • geographic variation,
               • time-scale of stress, and
               • ecological implications.

3:00 p.m.   BREAK

3:20 p.m.   Systematic identification of mitigation options (hard, soft, and
               preventive, including case study example of each if possible)
               relative to erosion processes already identified.

4:30 p.m.   BREAK

PLENARY SESSION #2—Maple Room

5:00 p.m.   Initial Report out from the three concurrent sessions

6:00 p.m.   Meeting adjourns for the day

WEDNESDAY, OCTOBER 5

Open Session

8:00 a.m.   BREAKFAST

PLENARY SESSION #3—Maple Room

8:30 a.m.   Meeting Reconvenes; summary of findings from geomorphic
              settings sessions—
              Jeff Benoit

9:00 a.m.   Brief overview of agenda for Wednesday breakouts—
              Jeff Benoit
              • identify outstanding issues from Day 1
              • goals for mitigation breakouts

9:10 a.m.   Overview of Mitigation Strategies—
              Don Ward, *U.S. Army Corps of Engineers*

9:30 a.m.   Public trust/property rights—
              Beth Bryant, *University of Washington, School of Marine Affairs*

9:50 a.m.   DISCUSSION

10:00 a.m.   BREAK

CONCURRENT SESSIONS—MITIGATION APPROACHES
              A. Manage Land Use
              B. Vegetate the Shoreline
              C. Harden the Shoreline
              D. Trap or Add Sediment

10:20 a.m.   Introductions and Purpose

10:30 a.m.   Background presentation by invited speaker
             Manage land use overview:
             Doug Myers, *Puget Sound Action Team*
             Vegetate the shoreline overview:
             Robin Lewis, *Lewis Environmental Services, Inc.*
             Harden the shoreline overview:
             Jay Tanski, *New York Sea Grant*
             Trap or add sediment overview: Fine Sediments,
             Phil Williams, *Phillip Williams Associates, Ltd.*; Coarse Sediments,
             Neville Reynolds, *Vanasse Hangen Brustlin, Inc.*

11:00 a.m.   Mini-presentations by willing participants

11:30 a.m.   Facilitated discussion of various techniques, technologies, and
             land management measures to mitigate erosion and inundation
             on sheltered coasts, including:
             • Current measure being used, with examples
             • Promising emerging approaches

Noon   LUNCH

1:00 p.m.   Facilitated discussion of effectiveness of various approaches
             from both engineering and ecological perspective, including:
             • Effectiveness relative to natural processes and human
               activities;
             • Identification of design features specific to natural or human
               causes of erosion;
             • Impacts, both individual and cumulative, on coastal resources
               (habitat, public/private property, and public access);
             • Timeframe required to assess effectiveness and impacts of
               various measures; and
             • Data necessary to predict when design criteria may be
               exceeded.

3:00 p.m.   BREAK

CLOSING PLENARY SESSION—Maple Room

3:45 p.m.   Summary and synthesis of findings from breakout groups

5:30 p.m.   Public workshop adjourns

THURSDAY, OCTOBER 6

Closed Session

8:00 a.m.   Committee work session

Open Session

Noon   LUNCH

1:00 p.m.   Field trip—advance registration required

PARTICIPANT LIST

Amanda Babson, *University of Washington*
Jeff Benoit, *SRA International*
Beth Bryant, *University of Washington, School of Marine Affairs*
Sarah Capote, *Ocean Studies Board, National Research Council*
Cyrilla Cook, *People for Puget Sound*
Scott Douglass, *University of South Alabama*
Lesley Ewing, *California Coastal Commission*
Monty Hampton, *U.S. Geological Survey, retired*
C. Scott Hardaway, Jr., *College of William and Mary, Virginia Institute of
    Marine Science*
Bernie Hargrave, *U.S. Army Corps of Engineers*
Debra Hernandez, *Hernandez and Company*
Robert Holman, *Oregon State University, College of Oceanic Atmospheric
    Sciences*
Joe Kelley, *University of Maine*
Evamaria Koch, *University of Maryland, Center for Environmental Science,
    Horn Point Laboratory*
Kathleen Kunz, *U.S. Army Corps of Engineers*
Jessica Lacy, *U.S. Geological Survey*
Robin Lewis, *Lewis Environmental Services, Inc.*
William Marsh, *University of British Columbia*
Neil McLellan, *Shiner Moseley and Associates*
Elliott Menashe, *Greenbelt Consulting*
Dan Miller, *New York State Department of Environmental Conservation, Hudson
    River National Estuarine Research Reserve (Representing CICEET)*
Andrew Morang, *U.S. Army Corps of Engineers*
Doug Myers, *Puget Sound Action Team*
James O'Connell, *Woods Hole Sea Grant*
Joan Oltman-Shay, *NorthWest Research Associates*

Jeff Parsons, *University of Washington*
Susan Peterson, *Teal Partners*
Denise Reed, *University of New Orleans, Department of Geology and Geophysics*
Neville Reynolds, *Vanasse Hangen Brustlin, Inc.*
Sue Roberts, *Ocean Studies Board, National Research Council*
Hugh Shipman, *Washington Department of Ecology*
Daniel Suman, *University of Miami, Rosenstiel School of Marine and Atmospheric Science*
Jay Tanski, *New York Sea Grant*
John Teal, *Teal Partners*
Ron Thom, *Battelle Marine Sciences Laboratory*
Jim Titus, *Environmental Protection Agency*
Heather Trim, *People for Puget Sound*
Don Ward, *U.S. Army Corps of Engineers*
Phil Williams, *Phillip Williams Associates, Ltd.*
Chin Wu, *University of Wisconsin*

D

Potential Federal Regulatory Requirements

A shoreline protection activity that requires a Section 404 permit from the USACE/USEPA may potentially require additional federal permits. A short description of potential federal requirements follows.

An individual permit application is subject to the provisions of the **National Environmental Policy Act (NEPA)** and may require preparation of an Environmental Impact Statement (EIS) (42 U.S.C. sec. 4332(2)(C)) or Environmental Assessment (EA) that the permit applicant finances and usually contracts to an environmental consultant. Preparation of an EIS may require a year while an EA may entail 3 months of effort and consultations.

If the State has an approved Coastal Management Program under the **Coastal Zone Management Act** (33 C.F.R. sec. 320.4(h)), the USACE will not grant a CWA sec. 404 permit to a non-federal applicant until certification of consistency with the State Program exists. In this case, the applicant must certify in the federal application that the proposed shoreline protection activity complies with the enforceable policies of the State Coastal Management Program and is consistent with the Coastal Management Plan. The State should notify the federal agency regarding its concurrence with or objection to the certification. No federal permit can be granted if the State objects to the applicant's certification, unless the Secretary of Commerce overrides the State's objection (16 U.S.C. sec. 1456(c)(3)(A)).

An additional requirement of the **Clean Water Act** must also be satisfied by the granting of a federal permit (33 U.S.C. sec. 1341(a)(1)). If the activity results in a discharge into navigable waters, the applicant must provide a Section

401 Water Quality Certification that the discharge complies with state effluent standards and any other water quality standards of the CWA.

The **Fish and Wildlife Coordination Act (FWCA)** (16 U.S.C. secs. 662-667e) requires coordination between the Fish and Wildlife Service (FWS) and the State fish and wildlife agencies where waters are controlled or modified through a federal permit. The purpose of this coordination is to prevent loss of or damage to wildlife. Consultation occurs between the federal and state agencies concerning mitigation of project-related losses of fish and wildlife resources. The FWCA is administered through the FWS and the NOAA's National Marine Fisheries Service.

While some provisions of the **Endangered Species Act (ESA)** apply to individuals on their private property, other provisions are applicable to federal agencies and have a different geographical scope. ESA Section 7 requires that all federal agencies insure that their actions do not jeopardize the continued existence of listed endangered or threatened species (16 U.S.C. sec. 1536(a)(2)). Thus, federal permits for shoreline erosion construction projects may be "federal agency actions" that may require formal or informal consultation between the USACE and the FWS or NOAA Fisheries, depending on the listed species. Some of the listed species that may require review for shoreline erosion control projects appear in Table D-1.

If the USACE determines after informal consultation that the proposed action is not likely to affect listed species or their critical habitat, and the FWS also agrees, then the review process terminates, and the permit is granted. However, if the FWS determines the proposed action may affect listed species and critical

TABLE D-1 Species Requiring Review for Shoreline Erosion Control Projects

Common Name of Listed Species	Scientific Name of Listed Species
Chinook salmon	*Oncorhynchus tshawytscha*
Sockeye salmon	*Oncorhynchus nerka*
Atlantic salmon	*Salmo salar*
Gulf sturgeon	*Acipenser oxyrinchus desotoi*
Shortnose sturgeon	*Acipenser brevirostrum*
Steelhead trout	*Oncorhynchus mykiss*
Green turtle	*Chelonia mydas*
Hawksbill turtle	*Eretmochelys imbricate*
Kemp's ridley turtle	*Lepidochelys kempii*
Leatherback turtle	*Dermochelys coriacea*
Loggerhead turtle	*Caretta caretta*
Olive ridley turtle	*Lepidochelys olivacea*
West Indian manatee	*Trichechus manatus*
Johnson Eye seagrass	*Halophila johnsonii*

habitat, then formal consultation begins, and the FWS/NOAA Fisheries must prepare a Biological Opinion that analyzes the effects of the action on the listed species and determines whether it will pose a threat of jeopardy to the continued existence of the species. The FWS' Biological Opinion may suggest alternatives to the project that will not cause jeopardy.

The federal permit requirement may also trigger compliance with the **National Historic Preservation Act (NHPA)** (16 U.S.C. sec. 470 *et seq.*). Section 106 of the NHPA requires that federal agency actions or permits consider their impact on any site or object that is included in or eligible for inclusion on the National Register of Historic Places (16 U.S.C. sec. 470f).

The 1996 **Sustainable Fisheries Act** introduced several conservation concepts to U.S. fisheries management, including Essential Fish Habitat (EFH) (16 U.S.C. sec 1853a)(7)). Fishery Management Councils are required to identify the Essential Fish Habitat in their Fishery Management Plans. The provisions of this statute create an interagency consultation process (16 U.S.C. sec. 1855(b)(1)(D)) in which the agencies must consult with NOAA Fisheries regarding actions that they propose to fund, authorize, or undertake that may adversely impact the EFH (16 U.S.C. sec. 1855(b)(2)). NOAA then recommends actions that the responsible agency can adopt to conserve the EFH. However, the consulting agency is not required to follow NOAA's recommendations.

E

Glossary

Accretion—the deposition of sediment, sometimes indicated by the seaward advance of a shoreline indicator such as the water line, the berm crest, or the vegetation line.

Armoring—the placement of fixed engineering structures, typically rock or concrete, on or along the shoreline to reduce coastal erosion. Armoring structures include seawalls, revetments, bulkheads, and riprap (loose boulders).

Beach—an accumulation of loose sediment (usually sand or gravel) along the coast.

Beach nourishment—the addition of sand (sand fill) to a shoreline to enhance or create a beach area, offering both shore protection and recreational opportunities.

Berm—a geomorphological feature usually located at mid-beach and characterized by a sharp break in slope, separating the flatter backshore from the seaward-sloping foreshore.

Bluff—an elevated landform composed of partially consolidated and unconsolidated sediments, typically sands, gravel and/or clays.

Breakwater— a single structure or a series of units placed offshore of the project site to reduce wave action on the shoreline.

Bulkhead—vertical shoreline stabilization structure that primarily retains soil.

Convergence—zones where sediment deposition exceeds sediment loss and accretion of sediment occurs.

Cumulative impacts—the impacts on the environment, which result from the incremental impact of the action when added to other past, present and reasonably foreseeable future actions regardless of what agency (federal or nonfederal) or person undertakes such other actions (40 CFR 1508.7 and 1508.8).

Delta—a nearly flat plain of alluvial deposit between diverging branches of the mouth of a river, often, though not necessarily, triangular.

Deposition—the process of sediment settling back to the bed or particles settling out of the water column.

Divergence—zones where the amount of sediment mobilized and lost exceeds the amount deposited.

Downdrift—in the direction of net longshore sediment transport.

Dune—a landform characterized by an accumulation of wind-blown sand, often vegetated.

Entrainment—the picking up and setting into motion of particles, either by wind or by water. The main entrainment forces are provided by impact, life force, and turbulence.

Erosion—the loss of sediment, sometimes indicated by the landward retreat of a shoreline indicator such as the water line, the berm crest, or the vegetation line. The loss occurs when sediment grains are entrained into the water column and transported away from the source.

Erosion mitigation—efforts to reduce or lessen the severity of *erosion*.

Eustatic sea-level rise—results from changes in global sea level. Eustatic changes represent global sea level. The causes can be complex, such as ice sheet melting, increasing temperature of the surface waters, increasing the volume of the spreading ridge.

Fetch—the distance that a wave travels from the point of origin to the shore where it breaks. In sheltered areas, the fetch corresponds to the distance across a span of water over which wind-generated waves may grow before breaking on the opposing shore.

Groins—a breakwater running seawards from the land, used on a variety of coasts including sheltered shores and open coasts, constructed to trap sand by interrupting *longshore transport*. A groin extends from the backshore into the *littoral zone* and is normally constructed perpendicular to the shore out of concrete, timbers, steel, or rock.

Hardening—see Armoring.

Infauna—animals that live in sediment.

Inundation—the temporary submergence of typically dry lands when there is an exceptional rise of the sea surface, and floodwaters cover the adjacent low-lying land.

Littoral cell—sections of coast for which sediment transport processes can be isolated from the adjacent coast. Within each littoral cell, a sediment budget can be defined that describes sinks, sources, and internal fluxes (sediment transport).

Littoral zone—used as a general term for the coastal zone influenced by wave action, or, more specifically, the shore zone between the high and low water marks (USACE, 2002b).

Longshore transport—sediment transport down the beach (parallel to the shoreline) caused by longshore currents and/or waves approaching obliquely to the shoreline.

Marsh—a vegetated mudflat.

Mudflat—an intertidal area with relatively fine sediment that may be vegetated by plant communities (marshes and mangroves) or colonized by microscopic plant communities (microalgae) and bacteria.

Offshore—the portion of the littoral system that is always submerged.

Open coast—tidal shores that have little or no protection from wave action.

Planform—the outline or shape of a body of water as determined by the still-water line (USACE, 2002b).

Pore water—water filling the spaces between grains of sediment.

Reach—a straight section of waterway that is uniform with respect to discharge, slope, and cross-section (USACE, 2002b).

Relative sea-level rise—the sea level relative to the land, which relates changes in local water levels to local land elevations. The rate of sea-level rise relative to a particular coast has a practical importance because some coastal land areas are subsiding, resulting in a relative rise in sea level, while other land areas are rising, resulting in a slower or falling sea level.

Revetment—a type of shoreline armoring that hardens the slope face and is often constructed from large boulders. A revetment tends to have a rougher (less reflective) surface than a seawall, and often is constructed with one or more layers of graded riprap but can also be constructed with precast concrete mats, timber, gabions (stone-filled, wire-mesh baskets), and other materials.

Scarp—a steep slope, usually along the foreshore and/or at the vegetation line, formed by wave attack.

Seawall—a vertical or near-vertical type of shoreline armoring characterized by a smooth surface. It retains soil, and reflects wave energy.

Sheltered coast—a coastal area sheltered by headlands, coves, natural or harbor breakwaters, tidal inlets, and river mouths and estuaries which have a limited distance between banks (fetch) and hence limited exposure to wind-driven waves. This area is usually characterized by low wave energies and stresses. These lower energy conditions foster habitats and ecological communities, such as marshes and mudflats, not typically found on open coasts.

Shore zone—the active volume of sediment affected by wave action.

Sill—a generally semicontinuous structure (e.g., a barrier of rock) built to reduce wave action and preserve, enhance, or create a marsh grass fringe for shore erosion control.

Storm surge—a temporary rise in sea level associated with a storm's low barometric pressure and onshore winds.

Wave attenuation—the power loss (the reduction in power density of a wave with distance) when a wave disperses over a larger area.

Wind fetch—the distance the wind blows over water with similar speed and direction.

SOURCE: Most terms are defined by the authors of this report. Any additional sources are listed at the end of each item.